The Lattice Dynamics and Statics of Alkali Halide Crystals

The Lattice Dynamics and Statics of Alkali Halide Crystals

John R. Hardy
University of Nebraska
Lincoln, Nebraska

and

Arnold M. Karo
Lawrence Livermore Laboratory
Livermore, California

Plenum Press · New York and London

Library of Congress Cataloging in Publication Data

Hardy, John R
   The lattice dynamics and statics of alkali halide crystals.

   Includes index.
   1. Alkali halide crystals. 2. Crystal lattices. I. Karo, Arnold M., joint author.
II. Title.
QD921.H335   548'.81   79-339
ISBN-13: 978-1-4613-2978-7   e-ISBN-13: 978-1-4613-2976-3
DOI: 10.1007/ 978-1-4613-2976-3

© 1979 Plenum Press, New York
Softcover reprint of the hardcover 1st edition 1979
A Division of Plenum Publishing Corporation
227 West 17th Street, New York, N.Y. 10011

Preface

Lattice dynamics is a classic part of solid state physics and the alkali halide crystals are classic materials. Nearly every new technique in many-body theory has first been applied to lattice-dynamical problems, and much of our present understanding of the physics of real crystals has its origins in pioneering work, both experimental and theoretical, carried out between 1920 and 1950 on alkali halide systems.

The object of the present text is to present a unified coverage of that part of physics where these two areas overlap and to extend this coverage somewhat in order to include not merely the dynamical behavior of alkali halides but also their static behavior. Specifically, we discuss the manner in which these materials respond to the presence of point imperfections. The rationale for this extension is simple: mechanics includes both dynamics and statics and a text which discusses the former should also discuss the latter.

Two other unifying themes are also present; the data presented are largely the result of our long collaboration in this area, and the work is a partial history of the impact of digital computers on lattice dynamics, an impact which parallels their impact on the whole of solid state physics.

Since this work is largely an account of model calculations, we have stressed the use of the simplest possible model at each level of sophistication and its uniform application to the crystals discussed. Specifically, we have confined ourselves almost entirely to lattice models whose parameters can be determined from macroscopic data (e.g., elastic constants, dielectric data) without reference to measured phonon dispersion curves. Again this has been done in the interests of uniformity of treatment of the whole sequence of materials. There has been some shading of emphasis; specifically we have chosen to give less space to the cesium chloride structure crystals than to the rock-salt structure materials and to limit our discussion of anharmonic effects to the bare minimum necessary to introduce our computational work on infrared lattice absorption. The first choice was largely one of style, dictated by the fact that only 3 out of the 20 alkali halides have this structure under normal conditions. The second choice is a consequence of our deliberate decision not to go beyond the harmonic approximation; the infrared results are presented only because they represent another set of data derived from harmonic calculations.

In listing references we have generally followed a policy of citing only those which are directly pertinent to our discussion and thus necessary to make the text self-contained.

Finally it is worth stressing at this point something which is implicit in much of the main text. The ultimate reason that model calculations of the type we discuss are at their best for alkali halides is that these are materials for which one has an excellent physical description of the nature of their bonding. Specifically, the Born–Mayer model and its various refinements* are physically sensible and lead to cohesive energies very close to the experimental values. It is probably fair to say that nothing quite like this type of understanding exists for any other class of materials.

Acknowledgments

The number of persons who have aided us in the production of this book is so large that it is impossible, for reasons of space, to

* See M. P. Tosi in *Solid State Physics* (F. Seitz and D. Turnbull, eds.), Vol. 16, p. 1. Academic Press, New York (1964).

acknowledge each of them individually. However, we would like to express specific thanks to Professors S. S. Jaswal and R. J. Hardy of the University of Nebraska, and to Professor C. W. McCombie and Dr. M. J. Sangster of the University of Reading in England, all of whom have provided us with invaluable assistance both by discussion and correspondence.

For most able assistance in computation and computer code development we would like to thank Ira Morrison, Fred McMurphy, and Curtis Geertgens of Lawrence Livermore Laboratory.

Finally, for their efforts in preparing the final manuscript and illustrations, we would like to thank the staff of the Technical Information Department at Lawrence Livermore Laboratory, and specifically technical editor Wallace Clements for his most able guidance of the project through all the many stages between the initial draft and the final product.

This work was performed in part under the auspices of the U.S. Department of Energy under contract No. W-7405-Eng-48.

John R. Hardy
Arnold M. Karo

Contents

I

Introduction

1. Historical Background

The theory of the many-body problem is a field of study that dates back to the very earliest days of theoretical classical mechanics, starting with the work of Lagrange and Hamilton and remaining the subject of continuing activity throughout the nineteenth and early twentieth centuries. The development of modern solid-state theory is in many respects a particular example of this general type of study since a solid is obviously a particular example of a many-body system. In treating a solid, in fact, one has to consider essentially two types of many-body problem. The first type involves the dynamics of the valence electrons of the crystal and necessitates the study of the elementary excitations of this system, insofar as they can be defined. The second type, which is the concern of this volume, involves the dynamics of the nuclear motion. These two problems differ in one fundamental respect: the valence electrons are completely de-localized and free to move throughout the whole crystal, whereas the nuclei are, to a good approximation, localized about the perfect-lattice sites. It is this essential difference that enables the nuclear motion to be treated, at least formally, with considerably greater precision than is possible in the case of the electronic system. This is because the localization of the nuclei can be shown to lead to the existence of clearly defined elementary excitations, called "phonons," of the whole system.

The formal development of the theory of these elementary excitations in fact preceded the development of modern quantum theory since it was carried out during the years 1912–1914 by Born and Von Kármán.[1-4] However, the results of quantum theory are not directly relevant to the problem of determining the elementary excitation spectrum of the system of nuclei, and the only real modification imposed by quantum theory is the requirement that these excitations be quantized. The spectrum itself is not affected. The significance of the elementary excitation spectrum of the system of nuclei is that within the so-called harmonic approximation, whose basic assumption is that the amplitude of the nuclear vibrations is small in comparison with the average internuclear spacing, one can transform from a set of Cartesian displacements to a set of generalized coordinates that are linear combinations of these Cartesian displacements. The normal coordinates so defined have the crucial property of being decoupled from each other. By this transformation there is thus obtained a complete formal solution of a particular class of many-body problem. This is unusual, since in most cases it is not possible to obtain any such exact formal solution. Such a solution permits examining the exact nature of these normal modes for real crystals in much more detail. This is distinct from the situation in which this exact separation cannot be made. In the latter case the tendency is to take the simplest possible theoretical model that offers any hope whatsoever of exact or near-exact solution. Hence the primary topic of this volume is the detailed work that has been carried out on the excitation spectra of real crystals. We shall restrict ourselves to the alkali halide sequence of crystals.

There has been a great deal of theoretical work, using the pseudopotential approach,[5,6] on the frequency spectra of metals. This has the advantage that it is close to being an *a priori* type of treatment of the many-body problem in that it starts, in principle, from the Schrödinger equation for the many-body system and seeks a solution for the normal-mode frequencies. However, the drawback to this approach is that the calculated phonon frequencies turn out to be differences between large quantities. Since the differences themselves are small, this results in a great sensitivity of the calculated phonon frequencies to the type of pseudopotential that is used. So far this type of approach has been applied almost exclusively to metals. Applications to insulating materials (e.g., silicon[7] and magnesium oxide[8]) have been made, but they have involved additional approximations in which "bond charges" are used to simulate the off-diagonal elements in the dielectric matrix.

In general, because of the extreme sensitivity of the calculated phonon frequencies to the effective electron–ion potential, a pseudopotential that reproduces the observed electronic band structure of any given material quite well does not necessarily reproduce its observed phonon frequencies with any degree of precision. Specifically, it appears that all such calculations are open to one type of criticism: the results are very sensitive to the Fourier transform of the screened pseudopotential at large values of the wave vector. This corresponds to a considerable uncertainty in the form of the direct-space potential at short distances, particularly at distances on the order of the first- and second-neighbor separation. A specific example is the case of the alkali metals, for which a number of authors[9–11] (see also Joshi and Rajagopal[6]) have constructed different screened pseudopotentials based on a variety of approximations both for the pseudopotential and the screening function. Although each of these will reproduce the observed dispersion curves fairly satisfactorily, the form of the potential at short distances can be very different for the various approximations. In particular, the position of the first minimum in the effective interionic potential can be markedly shifted with respect to the position of the first neighbor. This has a drastic effect on the values of the first and second derivatives of the interionic potential at the first-neighbor position. The existence of difficulties of this nature for even the simplest materials indicates a continuing need for semiempirical models which can be used to calculate phonon frequencies of more complex materials.

In this book, it is our concern to discuss extensively the results that have been obtained with various models for the alkali halide sequence of crystals. The models to be described are semiempirical, but, as we shall see, they also have a reasonably sound theoretical justification. Their main value is that they enable one to calculate all the allowed normal-mode frequencies for these crystals using very few parameters. These parameters can be seen to have a clearly defined physical significance and are determined from data other than the actual phonon frequencies. The success of these models varies with the alkali halide considered, and for this reason it is our intention to present systematic results for the whole sequence of crystals. We include in our discussion both the alkali halides that have the rock-salt structure and also the three cesium halides—the chloride, bromide, and iodide—that have the cesium chloride structure.

We present both calculations of the phonon frequencies them-selves and a discussion of the underlying principles of the various models, with the object of indicating the underlying theoretical justifications. We then present treatments of various properties that can be directly derived from the calculated phonon frequencies with the aid of certain additional assumptions. For example, it is our intention to discuss second-order Raman spectra and the side-band spectra of the fundamental infrared absorption of alkali halides; both types of spectra are intimately related to the calculated phonon frequencies and also involve the corresponding eigenvectors for the normal modes.

We also discuss thermal properties, such as the specific heat function, and at a later stage it is our intention to show the very close relationship that exists between calculations made on the dynamics of a crystal lattice and those concerned with the static properties of lattice defects (e.g., defect-formation energies and the displacement fields associated with such defects). We shall show that such calculations can be put in a form that is intimately related to that used in dealing with the dynamics of these crystals. This approach, which we refer to as lattice statics, has been exploited with marked success in dealing with defect properties.[12,13]

The alkali halides have been one of the most extensively studied types of solid both theoretically and experimentally. Attempts to understand theoretically their cohesive energies and similar properties go back to the earliest days of solid-state physics, specifically to the work of Born and his school beginning in 1912. Indeed, sodium chloride was one of the materials studied by Von Laue and co-workers by x-ray diffraction[14–16] and was thus one of the first substances definitely established to possess a periodic structure. This periodicity is the foundation of most modern solid-state physics. In particular, we shall exploit it extensively in our investigations of lattice dynamics.

The earliest model of these crystals, developed by Born and his school, regarded them as being made up of an assembly of point ions (plus and minus) that are prevented from collapsing under their mutual Coulomb attraction by the presence of a short-range repulsive interaction that in the simplest model is assumed to act only between nearest neighbors.[17–21] This picture has been singularly successful in accounting for the cohesive energies of these materials and has been given some quantum-mechanical justification by the work of Löwdin,[22] which we shall discuss in more detail later. However, this picture is not adequate, even from a semiclassical point of view, when one wishes to discuss the dynamical behavior of such materials. The reason for this is as follows: when the lattice is arbitrarily distorted, the various ions cease to be at sites of cubic symmetry, and therefore their charge distributions may acquire odd multipole moments, of which the most important is the dipole component. The crux of theoretical work on the lattice dynamics of these crystals is to take into account this electronic distortion.

The first extensive work on the dynamics of any ionic crystal was that carried out by Kellermann in 1940.[23] His treatment was based on the simple Born model, and he took into account the long-range nature of the Coulomb interaction in these materials and its effect on the lattice dynamics. (This Coulomb interaction does, of course, appear in the cohesive energy through the Madelung term.[17–21]) When the lattice is distorted, the displaced ions can be represented, to the lowest approximation, by dipoles placed at the perfect-lattice sites. There then appears in the energy change a dipole–dipole interaction summed over all sites in the crystal. It is not permissible to truncate this summation at any point since the interaction is of long range and the sum is only conditionally convergent. Thus the summation has to be carried out over the entire crystal, which Kellermann did by using the technique first proposed by Ewald[24,25] for the calculation of the Madelung energy, extended to the calculation of the dipole–dipole sum. In a later section we describe the technique in detail and give explicit results for the so-called Coulomb coupling coefficients derived by this means.

In addition to providing a means for actually calculating the Coulomb coupling coefficients for any type of lattice distortion, the technique provides the means of explicitly displaying the relationship between the lattice theory and macroscopic dielectric theory in the case of very-long-wavelength "optical" lattice vibrations. These are best regarded as extended molecular vibrations in which the positive and negative ions move in antiphase. However, it is crucial to the argument that, although it must be almost uniform over distances on the order of many interatomic spacings, the antiphase motion has nonetheless a modulating carrier wave whose wavelength is short in comparison with the macroscopic crystal dimensions. If this is not the case, the modes of the system become dependent on the geometry of the specimen being considered.

In all the foregoing arguments the electrostatic approximation is used; that is, the electromagnetic potentials are not retarded in any way, but are assumed to propagate with infinite velocity. This is not, of course, strictly correct, and it is necessary for very long wavelengths to take this retardation into account. We then enter what has become known as the "polariton" region, where the oscillations of the lattice and of the electromagnetic field of free space are coupled and the vibrations of the system can no longer be classified as either purely mechanical or purely electromagnetic, but have a mixed character. This has been extensively discussed by Huang,[26,27] and we shall give an account of this work. However, its effect on the lattice properties that we are concerned with is usually of no significance for most practical purposes.

In the lattice-statics part of the book, we shall have occasion to evaluate what we refer to as the "generalized force" exerted by a charged defect on the rest of the lattice, and here again we are dealing with a long-range Coulomb interaction that we handle by means of the theta-function transformation. Again the use of this technique enables us to establish the correspondence between our discrete lattice theory and the dielectric continuum theory, which has been used by earlier authors and which is the basis of the Fröhlich Hamiltonian[28] used for the polaron problem. We shall show that the theoretical work carried out with the Fröhlich Hamiltonian, motivated by its simplicity, has only limited validity. This is because the Fröhlich Hamiltonian is adequate only in treating the coupling of an extraneous charge with very-long-wavelength polar lattice displacements.

It is our hope that our present work will provide at a later stage the means of formulating the polaron Hamiltonian on a microscopic basis. However, this type of development is not one that is immediately practicable since the most we are capable of handling at present is a static charged defect, in which case there is a clearly defined center for the Coulomb force. In the case of the polaron, we are dealing with a charged particle that is free to migrate throughout the lattice. Thus the Coulomb force does not have a clearly defined center. This particular problem is also of importance in treating the theory of charged-defect migration through the lattice since here too one is dealing with a Coulomb force that does not have a clearly defined center—although in this type of problem there is some simplification in that, hopefully, only the energy of the system at a specific saddle-point configuration need be evaluated.

In this book we also attempt to give some treatment of defect vibrations in ionic crystals; that is, we consider the effect of introducing an impurity that has either a different mass or different interactions with its first neighbors. This type of problem has been treated very extensively elsewhere, particularly by Maradudin *et al.*,[29] and it is not our intention to give any detailed formal treatment. What we wish to do is to present some numerical results for the appropriate response functions that enter into defect-vibration calculations.

II

General Theory

2. Introduction of Normal Coordinates

In this section, it is our intention to show how it is possible to introduce a set of independent normal coordinates for a vibrating crystal lattice. Two basic assumptions are necessary:

1. The potential energy of the system can be expanded as a power series in the nuclear displacements.
2. It is adequate to truncate this series at the second-order term and to neglect terms of the third, fourth, and higher orders.

For crystals like the alkali halides this second assumption is reasonably valid, particularly at low temperatures. The basic criterion for its validity is that the crystal be relatively tightly bound; that is, the binding energy of the crystal must be very much greater than the zero-point vibrational energy of the nuclei. This is certainly true for the alkali halides, but it is by no means necessarily so for the rare-gas solids. The most extreme case is helium, which does not solidify at any temperature unless pressure is applied because the zero-point motion in this case is sufficiently large to disrupt the crystal. Similarly, neon, although it solidifies, has a relatively large zero-point energy, which makes it necessary to take into account the higher-order, or anharmonic, terms. However, here we shall work throughout within the harmonic approximation and therefore consider only the second-order term in the potential function. The reason for doing this is that within the harmonic approximation the problem of determining the independent normal coordinates is exactly soluble. The inclusion of higher-order terms in the displacements necessitates the use of approximations.

Our first assumption implies that there does exist a potential function for the nuclear motion—that is, that we are justified in writing the potential energy of the system as a function of the nuclear displacements alone and not explicitly considering the electronic degrees of freedom of the system. This procedure will be justified in a later section, but it should be mentioned at this stage that, when we proceed to consider specific models for the alkali halide crystals, we shall certainly find it necessary to allow for the polarizabilities of the ions in the crystal, and these polarizabilities and the associated electronic dipole moments constitute, in a sense, additional degrees of freedom. However, these additional degrees of freedom can be eliminated from the problem since the magnitudes of the electronic dipole moments are uniquely determined by the nuclear configuration at any given time.

In Eq. (2.1) we show the expression for the potential energy of the crystal expanded to second order in the displacements. We are considering a most general type of crystal lattice, which has the possibility of containing more than one type of atom within the primitive unit cell. It will be observed that there is no first-order term in the displacements in this series expansion because we are assuming that the expansion is made about the equilibrium position defined by the criterion that such first-order terms vanish. We have

$$U = U_0 + \frac{1}{2}\sum_{\substack{ll' \\ kk' \\ \alpha\beta}} \xi_\alpha\binom{l}{k}\left[\frac{\partial^2 U}{\partial \xi_\alpha\binom{l}{k}\,\partial \xi_\beta\binom{l'}{k'}}\right]_0 \xi_\beta\binom{l'}{k'} \qquad (2.1)$$

where the $\boldsymbol{\xi}$'s are the displacements, l and k index the primitive cell and the sublattice, respectively, α and β index the Cartesian components of the displacements, U is the total energy of the lattice, U_0 is its value for the undistorted lattice configuration, and the second term on the right-hand side is the lowest-order (second-order) nonvanishing term in the displacements.

Taking the dynamical part of the potential function in Eq. (2.1) and adding to it the kinetic energy of the nuclei, we obtain

$$H = \sum_{\substack{l \\ k \\ \alpha}} \frac{m_k}{2} \left[\dot{\xi}_\alpha \binom{l}{k} \right]^2 + \frac{1}{2} \sum_{\substack{ll' \\ kk' \\ \alpha\beta}} \xi_\alpha \binom{l}{k} \left[\frac{\partial^2 U}{\partial \xi_\alpha \binom{l}{k} \partial \xi_\beta \binom{l'}{k'}} \right]_0 \xi_\beta \binom{l'}{k'}$$

$$(2.2)$$

where m_k is the mass of the kth type of atom and a superposed dot denotes a first time derivative.

This constitutes the Hamiltonian function for the system. We now wish to obtain a linear transformation to a new set of coordinates such that the Hamiltonian given by Eq. (2.2) is reduced to a set of independent-harmonic-oscillator Hamiltonians.

In order to do this, we impose the so-called Born–Von Kármán periodic boundary conditions on our crystal. Hence we consider a large block of crystal that has the same shape as the primitive unit cell of the particular lattice we are considering. This supercell is assumed to have L unit cells in each of the directions of the three axes of the primitive unit cell and therefore contains $(L + 1)^3$ unit cells altogether. We then impose the boundary condition that equivalent points on opposite faces of this supercell have the same displacements. This means that we effectively reduce the number of degrees of freedom of the lattice to $3L^3 = 3N$ per sublattice.

This choice of boundary conditions is somewhat artificial, but it has the advantage, as we shall see later, of allowing us to use traveling-wave normal coordinates for our system and avoids the complications inherent in trying to use more realistic boundary conditions. The use of these oversimplified boundary conditions has been justified by Ledermann,[30]‡ who demonstrated that, for a densely distributed set of eigenfrequencies, the distribution function is influenced by whatever boundary conditions are used only to order $1/L$ (i.e., the number of surface atoms divided by the total number of atoms in the crystal).

We shall discuss this point in more detail later since Ledermann's argument assumes that the interatomic forces are of short range; however, it can be shown that, even when one has long-range forces, the argument is still valid except for a very small fraction (again, $\approx 1/L$) of particular normal coordinates. Thus we now take the most important step of our transformation—we write the ionic displacements as Fourier series, introducing complex normal coordinates as follows:

$$\boldsymbol{\xi}\begin{pmatrix} l \\ k \end{pmatrix} = (Nm_k)^{-1/2} \sum_{\mathbf{q}} \mathbf{W}(\mathbf{q}, k) \exp\left[2\pi i \mathbf{q} \cdot \mathbf{r}\begin{pmatrix} l \\ k \end{pmatrix}\right] \qquad (2.3)$$

where $\mathbf{r}\begin{pmatrix} l \\ k \end{pmatrix} = \mathbf{a}^l + \mathbf{r}\begin{pmatrix} 0 \\ k \end{pmatrix}$ is the position vector of ion $\begin{pmatrix} l \\ k \end{pmatrix}$ in the undistorted lattice.

The wave vectors in the Fourier series are restricted by our periodic boundary conditions to the following form:

$$\mathbf{q} = (1/L)(S_1 \mathbf{b}^1, S_2 \mathbf{b}^2, S_3 \mathbf{b}^3) \qquad (2.4)$$

where the $\mathbf{b}$ vectors are the basic vectors of the reciprocal lattice, whose general vector is denoted by $\mathbf{b}_h$, and S_1, S_2, S_3 are integers, $0 \leq S \leq L$. The $\mathbf{b}$ vectors are defined by

$$\mathbf{b}^\alpha \cdot \mathbf{a}_\beta = \delta_{\alpha\beta} \qquad (2.4a)$$

where the $\mathbf{a}$ vectors are the basic vectors of the direct space lattice.

‡ See also Born and Huang.[31]

Moreover, there is not an infinite number of allowed wave vectors since, if one considers all the wave vectors inside the first cell of reciprocal space, any wave vector that lies outside this cell simply gives a set of displacements at the lattice sites that are identical with those associated with a wave vector inside the first cell.

For convenience, it is customary to use this redundancy to fold back this first cell into a more symmetric region about the origin, known as the first Brillouin zone. In order to make this clear, Fig. 1 shows the primitive cells of the face-centered-cubic and simple-cubic lattices; Fig. 2 shows the associated first Brillouin zones. We have indicated on these figures the various symmetry points and symmetry axes that we shall have occasion to refer to later (the notation follows that of Bouckaerdt, Smoluchowski, and Wigner[32]).

At this stage, we wish to establish a mathematical lemma that we shall have occasion to use in the course of transforming our Hamiltonian to the independent-harmonic-oscillator Hamiltonians.

Consider the sum

$$N^{-1} \sum_{\substack{l \\ \text{supercell}}} \exp(2\pi i \cdot \mathbf{q} \cdot \mathbf{a}^l)$$

$$= N^{-1} \sum_0^L \exp\left(2\pi i \frac{S_1 l_1}{L}\right) \sum_0^L \exp\left(2\pi i \frac{S_2 l_2}{L}\right) \sum_0^L \exp\left(2\pi i \frac{S_3 l_3}{L}\right)$$

$$= N^{-1}\left[\frac{1 - \exp(2\pi i S_1)}{1 - \exp(2\pi i S_1/L)}\right]\left[\frac{1 - \exp(2\pi i S_2)}{1 - \exp(2\pi i S_2/L)}\right]$$

$$\times \left[\frac{1 - \exp(2\pi i S_3)}{1 - \exp(2\pi i S_3/L)}\right] \tag{2.5}$$

Since S_1, S_2, S_3 are integers, this expression sums to zero unless each S value is an integral multiple of L; that is, unless $\mathbf{q}$ is a reciprocal lattice vector. In this case, direct summation shows that

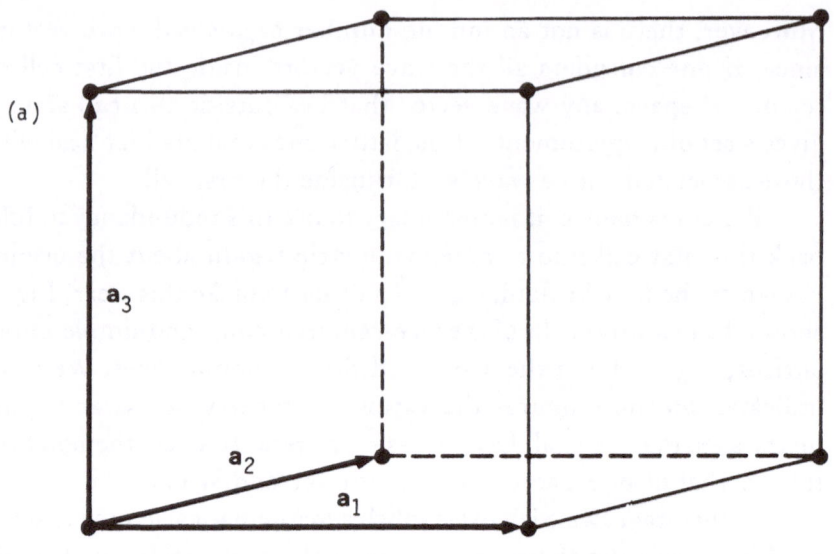

(a)

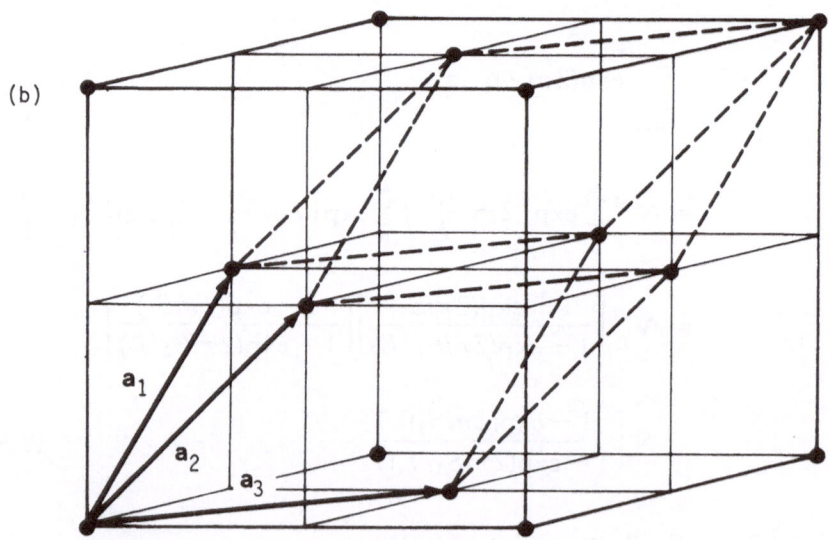

(b)

FIG. 1. Crystallographic and primitive unit cells for (a) simple-cubic (CsCl structure) and (b) face-centered-cubic (NaCl structure) lattices.

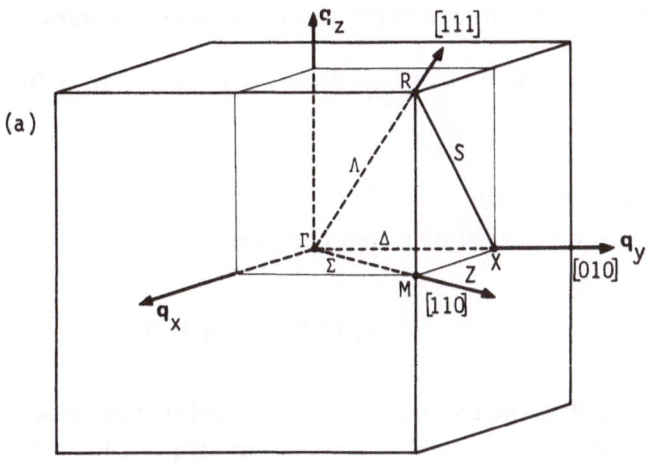

(a)

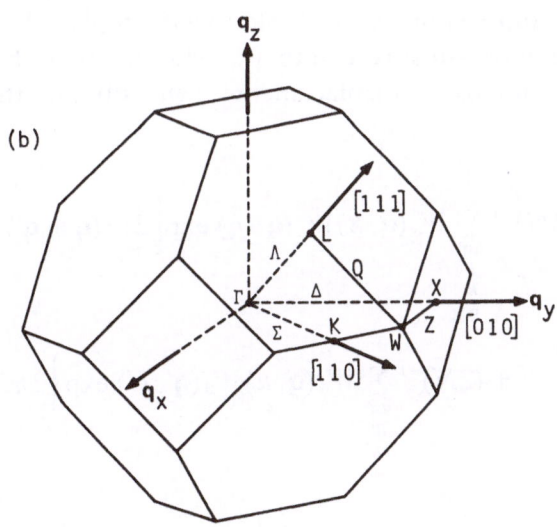

(b)

FIG. 2. Brillouin zones for (a) simple-cubic and (b) face-centered-cubic lattices.

the sum is unity. By a similar argument, one can show that

$$N^{-1} \sum_{\substack{\mathbf{q} \\ \text{Brillouin} \\ \text{zone}}} \exp(2\pi i \mathbf{q} \cdot \mathbf{a}') = 0 \qquad (\mathbf{q} \neq 0)$$

unless $\mathbf{a}' = (0, 0, 0)$.

Thus we have shown that the sum

$$N^{-1} \sum_{l} \exp[2\pi i (\mathbf{q}' + \mathbf{q}'') \cdot \mathbf{a}']$$

vanishes unless the two wave vectors $\mathbf{q}'$ and $\mathbf{q}''$ add up to some vector of the reciprocal lattice as defined by Eq. (2.4a). Moreover the restrictions on $\mathbf{q}'$ and $\mathbf{q}''$ are such that the origin is the only possibility.

The first Brillouin zone is equivalent to the first cell of this lattice, and it can be seen that the reciprocal of the face-centered-cubic lattice is body centered; and, as one would expect, the reciprocal of the simple-cubic lattice is also itself simple cubic.

If we now substitute into the Hamiltonian [Eq. (2.2)] the Fourier-decomposed displacements, we obtain the following result:

$$H = (2N)^{-1} \sum_{\substack{qq' \\ l \\ k \\ \alpha}} \dot{W}_\alpha(\mathbf{q}, k) \dot{W}_\alpha(\mathbf{q}', k) \exp\left[2\pi i (\mathbf{q} + \mathbf{q}') \cdot \mathbf{r}\begin{pmatrix} l \\ k \end{pmatrix}\right]$$

$$+ (2N)^{-1} \sum_{\substack{qq' \\ ll' \\ kk' \\ \alpha\beta}} W_\alpha(\mathbf{q}, k) W_\beta(\mathbf{q}', k') \exp\left\{2\pi i \left[\mathbf{q} \cdot \mathbf{r}\begin{pmatrix} l \\ k \end{pmatrix}\right.\right.$$

$$\left.\left. + \mathbf{q}' \cdot \mathbf{r}\begin{pmatrix} l' \\ k' \end{pmatrix}\right]\right\} U\begin{pmatrix} ll' \\ kk' \\ \alpha\beta \end{pmatrix} \qquad (2.6)$$

where

$$U\begin{pmatrix} ll' \\ kk' \\ \alpha\beta \end{pmatrix} = (m_k m_{k'})^{-1/2}\left[\partial^2 U \Big/ \partial\xi_\alpha\begin{pmatrix} l \\ k \end{pmatrix} \partial\xi_\beta\begin{pmatrix} l' \\ k' \end{pmatrix}\right]_0 \qquad (2.6a)$$

After a certain amount of manipulation and by using the lemma just established, we find that Eq. (2.6) becomes

$$H = \tfrac{1}{2}\sum_{\substack{q \\ k \\ \alpha}} \dot{W}_\alpha^*(\mathbf{q}, k)\dot{W}_\alpha(\mathbf{q}, k) + \tfrac{1}{2}\sum_{\substack{q \\ kk' \\ \alpha\beta}} W_\alpha^*(\mathbf{q}, k)U\begin{pmatrix} \mathbf{q} \\ kk' \\ \alpha\beta \end{pmatrix} W_\beta(\mathbf{q}, k')$$

$$(2.7)$$

The matrix

$$U\begin{pmatrix} \mathbf{q} \\ kk' \\ \alpha\beta \end{pmatrix} = (m_k m_{k'})^{-1/2}\left[\partial^2 U \Big/ \partial\xi_\alpha\begin{pmatrix} l \\ k \end{pmatrix} \partial\xi_\beta\begin{pmatrix} l' \\ k' \end{pmatrix}\right]_0$$

$$\times \exp\left\{2\pi i\mathbf{q} \cdot \left[\mathbf{r}\begin{pmatrix} l \\ k \end{pmatrix} - \mathbf{r}\begin{pmatrix} l' \\ k' \end{pmatrix}\right]\right\} \qquad (2.8)$$

which is the Fourier transform of the mass-weighted force-constant matrix [Eq. (2.6a)], is referred to as the dynamical matrix. It is a $3n \times 3n$ matrix, n being the number of atoms in the basis. Thus in the case of the alkali halides it will be a 6×6 matrix.

We have now almost completed the transformation of the original Hamiltonian; we have not completely eliminated cross terms, but we have reduced it from a sum of $3Nn \times 3Nn$ terms to a sum of $N(3n \times 3n)$ terms.

The final stage of the diagonalization is straightforward. We take the dynamical matrix and we look for its eigenvalues and their associated eigenvectors, given by

$$\sum_{k'\beta} U\begin{pmatrix} \mathbf{q} \\ kk' \\ \alpha\beta \end{pmatrix} \left[e_\beta\left(k' \middle| \begin{matrix} \mathbf{q} \\ j \end{matrix}\right) \right] = \omega^2\left(\begin{matrix} \mathbf{q} \\ j \end{matrix}\right) e_\alpha\left(k \middle| \begin{matrix} \mathbf{q} \\ j \end{matrix}\right) \qquad (2.9)$$

The eigenvectors $\mathbf{e}(k \middle| \begin{smallmatrix} \mathbf{q} \\ j \end{smallmatrix})$ defined by these equations satisfy both the direct and inverse orthonormal relations

$$\sum_{k\alpha} e_\alpha^*\left(k \middle| \begin{matrix} \mathbf{q} \\ j' \end{matrix}\right) e_\alpha\left(k \middle| \begin{matrix} \mathbf{q} \\ j \end{matrix}\right) = \delta_{jj'} \qquad (2.10a)$$

and

$$\sum_j e_\beta^*\left(k' \middle| \begin{matrix} \mathbf{q} \\ j \end{matrix}\right) e_\alpha\left(k \middle| \begin{matrix} \mathbf{q} \\ j \end{matrix}\right) = \delta_{kk'}\delta_{\alpha\beta} \qquad (2.10b)$$

In general Eqs. (2.9) and (2.10) will define the eigenvectors completely. However, if it happens that one or more of the eigenvalues $\omega^2(\begin{smallmatrix} \mathbf{q} \\ j \end{smallmatrix})$ are degenerate, one has infinite freedom in choosing the associated eigenvectors. This is of no consequence because it is always possible to choose one set satisfying the direct and inverse orthonormal relations, and this is all we require. If we now define generalized coordinates by

$$Q\left(\begin{matrix} \mathbf{q} \\ j \end{matrix}\right) = \sum_{k\alpha} e_\alpha^*\left(k \middle| \begin{matrix} \mathbf{q} \\ j \end{matrix}\right) W_\alpha(\mathbf{q}, k) \qquad (2.11)$$

and use Eqs. (2.9) and (2.10), the Hamiltonian becomes

$$\tfrac{1}{2}\sum_{\mathbf{q}j}\left[\dot{Q}^*\left(\begin{matrix} \mathbf{q} \\ j \end{matrix}\right)\dot{Q}\left(\begin{matrix} \mathbf{q} \\ j \end{matrix}\right) + \omega^2\left(\begin{matrix} \mathbf{q} \\ j \end{matrix}\right) Q^*\left(\begin{matrix} \mathbf{q} \\ j \end{matrix}\right) Q\left(\begin{matrix} \mathbf{q} \\ j \end{matrix}\right) \right] \qquad (2.12)$$

which is now a sum of independent-harmonic-oscillator Hamiltonians. Finally, for completeness, we rewrite this Hamiltonian using creation and destruction operators. This is done by realizing that the momentum canonically conjugate to $Q\binom{q}{j}$ is $\dot{Q}\binom{-q}{j}$. We obtain this result by using Lagrange's prescription for the canonical momentum:

$$\partial L / \partial \dot{Q}\binom{q}{j} = P\binom{q}{j}$$

Thus

$$\left[\dot{Q}\binom{-q}{j}, Q\binom{q}{j}\right] = i\hbar$$

and if we define the operators $a_{qj}^{\dagger}$ and a_{qj} by

$$Q\binom{q}{j} = \left[\hbar / 2\omega\binom{q}{j}\right]^{1/2} (a_{qj}^{\dagger} + a_{-qj})$$

and

$$Q\binom{-q}{j} = \left[\hbar / 2\omega\binom{q}{j}\right]^{1/2} (a_{-qj}^{\dagger} + a_{qj})$$

then, for the commutation relation between P and Q to hold, it is necessary that

$$[a_{qj}^{\dagger}, a_{q'j'}] = \delta_{qq'}\delta_{jj'}$$

and the total Hamiltonian becomes

$$\sum_{qi} \hbar\omega\binom{q}{j}\left[a^{\dagger}\binom{q}{j}a\binom{q}{j} + \frac{1}{2}\right] \tag{2.13}$$

This formal work has shown us how to diagonalize our initial Hamiltonian; however, for most practical purposes, this is not the way in which one actually determines the normal-mode frequencies. The standard method of doing this is to take the equation of motion for a given ion, which has the following form:

$$m_k \ddot{\xi}_\alpha \binom{l}{k} = -\sum_{\substack{l' \\ k'\beta}} \left[\partial^2 U \Big/ \partial \xi_\alpha \binom{l}{k} \partial \xi_\beta \binom{l'}{k'} \right]_0 \xi_\beta \binom{l'}{k'} \qquad (2.14)$$

Then we substitute into this a plane-wave solution,

$$\boldsymbol{\xi} \binom{l}{k} = \frac{\mathbf{W}(k)}{(m_k)^{1/2}} \exp\left\{ i \left[2\pi \mathbf{q} \cdot \mathbf{r} \binom{l}{k} - \omega t \right] \right\} \qquad (2.15)$$

with the normal imaginary exponential time dependence. This leads to the eigenvalue equations

$$\omega^2 W_\alpha(k) = \sum_{k'\beta} U \begin{pmatrix} \mathbf{q} \\ kk' \\ \alpha\beta \end{pmatrix} W_\beta(k') \qquad (2.16)$$

and it can readily be seen that the eigenvalues of this equation and the associated eigenvectors are exactly those that we have used in the foregoing transformation.

Our phase convention differs slightly from that used by Born and Huang[33] in their reduction of the Hamiltonian, in that we have used the same phase convention in our reduction as that employed in the equations of motion. The eigenvectors defined in this way are such that the long-wavelength acoustic vibrations (i.e., those in which the two types of ion in a given cell vibrate in phase) reduce to the standard sound waves that one obtains from the solution of the Riemann–Christoffel equations of continuum elasticity.

For a lattice containing n atoms in the unit cell, one obtains $3n$ normal modes for a given wave vector. For a diatomic crystal, this implies six normal modes. For small wave vectors, three of these are the normal acoustic vibrations, which can be identified with macroscopic sound waves; the other three have the property that the atoms in the unit cell are moving in antiphase. These are the so-called optic vibrations. This nomenclature has a simple origin: in an ionic crystal these vibrations have an associated dipole moment and can therefore couple to an external electromagnetic field. There is, however, an important difference between the character of such vibrations for ionic and nonionic crystals. For ionic materials, there is associated with the longitudinal optic type of vibration a macroscopic electric field that has the effect of raising the frequency of this type of vibration with respect to that of the transverse vibrations. Such an effect is absent in nonpolar crystals, and we shall discuss this at greater length in a later section, where our aim will be to show how the microscopic theory leads to this splitting and how it can be put into correspondence with the macroscopic theory of Huang.[26,27]

At this stage it should be stated explicitly that the reduction of our initial Hamiltonian to the final set of decoupled-harmonic-oscillator Hamiltonians has been made possible by exploiting the translational invariance of the crystal lattice. This can be seen from Eqs. (2.7) and (2.14). Use has been made of the fact that the ionic mass is site independent provided that one remains on the same sublattice. Similarly, use has been made of the fact that the force-constant matrix depends only on the difference between the coordinates of the two sites that one is considering, and not on their absolute positions in the crystal. One could object to this last assumption in the case of ions that are near the surface, but previous arguments justify our assuming that the effect of the free surface is on the order of the ratio of the number of surface sites to the number of interior sites. Thus, if one were dealing with specimens with a large surface-to-volume ratio, surface effects would necessitate corrections that would require detailed examination.

In principle, it is always possible to diagonalize a Hamiltonian of the form given in Eq. (2.2), but for a large system with no long-range symmetry, such as an amorphous material, it is not possible to derive analytical expressions for the proper transformation functions. In the sections on imperfect lattices we shall also make extensive use of the translational invariance of the perfect lattice.

3. The Adiabatic Approximation

In the preceding section we made implicit use of the adiabatic approximation when we assumed the existence of an appropriate potential function for the nuclear motion. It is our aim here to justify this assumption by presenting the Born–Oppenheimer treatment of the nuclear motion.[34,35] This is a very general treatment and can be applied to any molecular system, large or small. For our present work we shall use a simplified derivation that imposes more restrictive conditions than are truly necessary. It can be shown that the adiabatic approximation is good even for metals, but our present derivation is such that this is not apparently the case. The true expansion parameter is the ratio (electronic mass/nuclear mass)$^{1/4}$, which is obviously small, but in our present derivation we shall merely require that the gap in the electron-energy-level spectrum between the highest filled state and the first excited state be very much larger than the maximum phonon frequency.

The first step in this derivation is the assumption that one can write any wave function for the system in the following form:

$$\Psi(\xi, x) = \sum_n \chi_n(\xi)\phi_n(x, \xi) \tag{3.1}$$

where $\xi \equiv \xi_\alpha\binom{l}{k}$ for all nuclei, x denotes all electronic coordinates, and n labels the possible electronic states of the system.

The object of the derivation is to show that, if one chooses the basis functions appropriately, it is possible, to a good approximation, to decouple the various terms in this series into families of terms, each one of which is a set of vibrational levels that belongs to a single electronic state. However, the electronic state does have a certain implicit dependence on the nuclear configuration.

To do this, we write down the total Hamiltonian for the system of ions plus electrons. Thus we have

$$H = T_E + T_N + U(x, \xi) \qquad (3.2)$$

where T_E and T_N are, respectively, the electronic and nuclear kinetic energies and $U(x, \xi)$ is the total potential energy. Given this Hamiltonian operator, we now separate out the part that acts on the electronic coordinates. We assume that the functions $\phi_n(x, \xi)$ are solutions of the reduced Schrödinger equation

$$H^0 \phi_n(x, \xi) = [T_E + U(x, \xi)] \phi_n(x, \xi) = \phi_n(\xi) \phi_n(x, \xi) \qquad (3.3)$$

where $H^0 = T_E + U(x, \xi)$.

We now look for eigenfunctions of the whole Hamiltonian that are of the form given by Eq. (3.1). We are now in a position to decide on the appropriate choice of $\phi_n(x, \xi)$ by using Eqs. (3.1) and (3.2) and the total Schrödinger equation. Thus we obtain

$$[T_N + \phi_n(\xi) - E] \chi_n(\xi) + \sum_{n'} C_{nn'}(\xi, \mathbf{P}) \chi_{n'}(\xi) = 0 \qquad (3.4)$$

Here $\mathbf{P} \equiv \mathbf{P}\binom{l}{k}$ is the nuclear momentum operator, and we have premultiplied by $\phi_n^*(x, \xi)$ and integrated over x. The term $C_{nn'}$ is defined by

$$C_{nn'} = \sum_{\substack{l \\ k \\ \alpha}} m_k^{-1} \left[A_{nn'} \binom{l}{k}_\alpha P_\alpha \binom{l}{k} + B_{nn'} \binom{l}{k}_\alpha \right] \qquad (3.4a)$$

where

$$A_{nn'}\begin{pmatrix} l \\ k \end{pmatrix}_\alpha = \int \phi_n^*(x, \xi) P_\alpha \begin{pmatrix} l \\ k \end{pmatrix} \phi_{n'}(x, \xi)\, dx$$

and

$$B_{nn'}\begin{pmatrix} l \\ k \end{pmatrix}_\alpha = \frac{1}{2} \int \phi_n^*(x, \xi) P_\alpha^2 \begin{pmatrix} l \\ k \end{pmatrix} \phi_{n'}(x, \xi)\, dx$$

If all the coefficients $C_{nn'}$ were negligible, we would have solved the problem of decoupling the nuclear and electronic motions, and Eq. (3.4) would be the appropriate Schrödinger equation for the nuclear motion; to second order in the nuclear displacements, this assumes the coupled-harmonic-oscillator form.

However, we first have to justify the approximation of dropping the $C_{nn'}$ coefficients. This is not particularly easy to do with any degree of rigor. Let us consider first the diagonal terms (C_{nn}). Of the two contributions A_{nn} and B_{nn}, the second is the average of the nuclear kinetic energy, with respect to the electronic coordinates. This term is not necessarily small, but as a function of nuclear coordinates it is more or less constant since the dominant contribution comes from close to the nucleus and will be virtually independent of the lattice configuration. Thus we can regard this as a small correction to the energy of the ground state. Moreover, since we are considering the energy of a stationary state, the remaining term A_{nn} does not contribute since its expectation value is effectively the rate of change of the normalization with respect to ionic configuration, and this is certainly zero if the wave functions are chosen to be real. (This may be done without loss of generality.)

The off-diagonal terms $C_{nn'}$ lead to the coupling of the electronic and nuclear motions. We shall neglect these nonadiabatic terms since they give a correction to the energy that is certainly small for ionic crystals owing to the large band gap, as can be shown by perturbation theory. However, they are present in principle. For example, they are responsible for the electron–phonon scattering in metals. The standard method of treating these terms is to drop the rather cumbersome expression in Eq. (3.4) and to replace it by an effective electron–phonon interaction, which is usually computed with respect to the rigid-lattice electron eigenstates, that is, the Bloch states.

However, it is worth noting, since this point does not appear to be emphasized in the literature, that eliminating the linear term $A_{nn'}$ in a given electronic state is not something that can be done on a once-for-all basis. That is to say, the electronic wave functions that eliminate this interaction for a given state are not the same as those that will eliminate it for, say, the first excited state. Therefore, when an electron from the ground state is excited to the first excited state, there will be a linear electron–lattice coupling in this state. This coupling is particularly strong in the case of ionic crystals as a result of the macroscopic field associated with the long-wavelength optical modes. It leads, for example, to a strong broadening of the exciton lines observed in ionic crystals.

In fact, it would seem from this result that a phonon in an ionic crystal is a rather complicated excitation involving the virtual excitation of electrons from the ground state to excited states with phonon emission. Thus, in principle, a phonon is a first approximation to the solution of a coupled hierarchy of equations.

In practice the wave functions $\phi(x, \xi)$, which contain this implicit dependence on the nuclear configuration, are not themselves used. Instead the starting point is the rigid-lattice eigenfunctions, from which is obtained the change in energy resulting from lattice distortion and from that the effective nuclear potential function for a given electronic state. In a later section we shall describe how this is done for the various model calculations.

There is something of a gap between the Born–Oppenheimer arguments and those used in developing potential functions for the various model calculations. However, the Born–Oppenheimer approximation provides the basic justification for computing the effective potential function by the evaluation of the change in energy of the whole system for the static, but distorted, lattice. As Ziman puts it, the electrons are treated as "slaves" that instantaneously follow the motion of the nuclei. However, there is always a residual electron–phonon interaction that cannot be eliminated but can be neglected to a good approximation. A determination of the exact form of this interaction would appear to be rather complicated, but it does not seem to us that one is justified in using the type of effective electron–phonon interaction that is used in, for example, semiconductor mobility calculations. There one employs a deformation potential.[36] Alternatively, in polar materials one employs the Fröhlich interaction. These describe the interaction of the lattice with an *additional* electron that has been introduced into the system in the first available excited state. In other words, the deformation potential[36] or the Fröhlich interaction[28] corresponds to the linear terms in the electron–phonon interaction, which can be eliminated from the ground-state Hamiltonian by perturbation theory. Therefore they do not appear to be the appropriate forms to use for computing the residual electron–phonon interaction within the adiabatic approximation.

We now have a formal justification for the adiabatic approximation, which we shall employ in our subsequent calculations with regard to the development of the model potential functions to be used in our lattice-dynamics calculations for the alkali halides.

III

Dipolar Models

4. Long-Wave Optical Vibrations of Cubic Ionic Lattices

In this section we propose to describe the effects that are peculiar to the coupled electromagnetic-field–lattice-oscillator system. In this regime, which is of importance only when the finite wavelength of **E** and **H** cannot be ignored, the transverse normal modes of the system assume a "mixed" character, being neither purely mechanical oscillations nor pure light waves. These hybrid modes have been given the name "polaritons" and have been observed by Raman scattering,[37,38] which confirmed the basic ideas developed by Huang.[26,27] However, to observe these excitations in this way, one has to use a noncentrosymmetric material such as gallium phosphide. The present theory is exact for such zinc-blende structures.

4.1. Macroscopic Theory

Our approach follows Huang[26,27,39] and is purely classical. First we assume that there are no free charges and that we are dealing with an isotropic dielectric.

Thus we have the four Maxwell equations

$$\mathbf{\nabla} \cdot \mathbf{D} = 0 \tag{4.1}$$

$$\mathbf{\nabla} \cdot \mathbf{H} = 0 \qquad (\mu = 1 \text{ assumed}) \tag{4.2}$$

$$\mathbf{\nabla} \times \mathbf{E} = -c^{-1}\dot{\mathbf{H}} \tag{4.3}$$

$$\mathbf{\nabla} \times \mathbf{H} = c^{-1}(\dot{\mathbf{E}} + 4\pi\dot{\mathbf{P}}) \tag{4.4}$$

These equations have a clear meaning in the region of interest since the wavelengths of concern are substantially greater than the lattice constant, and quantities such as $\mathbf{D}$ are well-defined spatial averages over many lattice cells. To complement these four equations we have another two that we first postulate and then justify.

First, we must have a "macroscopic" optical sublattice displacement (i.e., some average over many cells of a microscopic displacement). For this we use the vector

$$\mathbf{W} = \left[\frac{M_+ M_-}{(M_+ + M_-)v_a} \right]^{1/2} (\mathbf{u}_+ - \mathbf{u}_-)$$

where the M terms are the ionic masses, the $\mathbf{u}$ terms are the corresponding average sublattice displacements taken over many cells, and v_a is the unit-cell volume. Now we can write

$$\ddot{\mathbf{W}} = b_{11}\mathbf{W} + b_{12}\mathbf{E} \tag{4.5}$$

and

$$\mathbf{P} = b_{21}\mathbf{W} + b_{22}\mathbf{E} \tag{4.6}$$

For the present, these are postulated; moreover, the conservation of energy requires that $b_{12} = b_{21}$, although this, too, will later be proved directly.

Now let us consider the detailed derivation of the polariton modes. We can write

$$\mathbf{W} = \mathbf{W}_0 \exp[i(\mathbf{k} \cdot \mathbf{r} - \omega t)]$$

$$\mathbf{P} = \mathbf{P}_0 \exp[i(\mathbf{k} \cdot \mathbf{r} - \omega t)]$$

$$\mathbf{E} = \mathbf{E}_0 \exp[i(\mathbf{k} \cdot \mathbf{r} - \omega t)]$$

$$\mathbf{H} = \mathbf{H}_0 \exp[i(\mathbf{k} \cdot \mathbf{r} - \omega t)]$$

since we have linear equations and the solutions superpose; thus we can consider each Fourier component separately.

If we substitute these solutions into Eqs. (4.5) and (4.6), we obtain

$$- \omega^2 \mathbf{W} = b_{11}\mathbf{W} + b_{12}\mathbf{E} \tag{4.7}$$

$$\mathbf{P} = b_{21}\mathbf{W} + b_{22}\mathbf{E} \tag{4.8}$$

Normally one writes the dielectric constant without damping as

$$\varepsilon = \varepsilon_\infty + \frac{\varepsilon_0 - \varepsilon_\infty}{1 - (\omega/\omega_0)^2} \tag{4.8a}$$

where ε_0 is the *total* static dielectric constant, ε_∞ is the electronic component, and ω_0 is the dispersion frequency.

Thus we have, at once, the relations

$$b_{11} = -\omega_0^2$$

$$b_{12} = b_{21} = \left(\frac{\varepsilon_0 - \varepsilon_\infty}{4\pi}\right)^{1/2} \omega_0$$

$$b_{22} = (\varepsilon_\infty - 1)/4\pi$$

$$\mathbf{k} \cdot (\mathbf{E} + 4\pi\mathbf{P}) = 0 \tag{4.9}$$

$$\mathbf{k} \cdot \mathbf{H} = 0 \tag{4.10}$$

$$\mathbf{k} \times \mathbf{E} = (\omega/c)\mathbf{H} \tag{4.11}$$

$$\mathbf{k} \times \mathbf{H} = (-\omega/c)(\mathbf{E} + 4\pi\mathbf{P}) \tag{4.12}$$

Equation (4.7) gives

$$\mathbf{W} = b_{12}\mathbf{E}/(-b_{11} - \omega^2) \tag{4.13}$$

(note that there is no damping), and thus

$$\mathbf{P} = \left[\frac{b_{12}^2}{(-b_{11} - \omega^2)} + b_{22}\right]\mathbf{E} \tag{4.14}$$

Thus, on substituting into Eq. (4.9), we have

$$\left[1 + \frac{4\pi b_{12}^2}{(-b_{11} - \omega^2)} + 4\pi b_{22}\right]\mathbf{k} \cdot \mathbf{E} = 0 \tag{4.15}$$

This allows two classes of solutions: either case A,

$$1 + \frac{4\pi b_{12}^2}{(-b_{11} - \omega^2)} + 4\pi b_{22} = 0$$

or case B,

$$\mathbf{k} \cdot \mathbf{E} = 0$$

In case A we have, from Eq. (4.12),

$$\mathbf{k} \times \mathbf{H} = 0$$

as

$$\mathbf{E} + 4\pi \mathbf{P} = 0$$

and Eq. (4.10) gives

$$\mathbf{k} \cdot \mathbf{H} = 0$$

Thus, as **H** has no component either perpendicular or parallel to **k**,

$$\mathbf{H} = 0$$

and it follows from Eq. (4.11) that

$$\mathbf{k} \times \mathbf{E} = 0$$

Furthermore, since **E** is not equal to zero and **E** is parallel to **k**, we have, from Eqs. (4.13) and (4.14), longitudinal solutions

$$\mathbf{W} \| \mathbf{E} \| \mathbf{k} \| \mathbf{P}$$

whose frequency ω_l is given by

$$\varepsilon_\infty + \frac{4\pi b_{12}^2}{(-b_{11} - \omega_l^2)} = 0$$

On substituting for the b coefficients, we obtain

$$\omega_l^2 = (\varepsilon_0/\varepsilon_\infty)\omega_0^2$$

which is exactly the solution that is obtained by neglecting retardation.

In case B we have

$$\mathbf{k} \cdot \mathbf{E} = 0$$

that is, $\mathbf{E}$ is perpendicular to $\mathbf{k}$. Combining this result with Eqs. (4.10) and (4.11), we find that $\mathbf{H}$ is perpendicular to $\mathbf{k}$ and $\mathbf{E}$ and obtain the following equation:

$$kE = (\omega/c)H \tag{4.16}$$

since Eq. (4.11) can be written in scalar form. Similarly, Eq. (4.12) also becomes a scalar equation:

$$kH = (\omega/c)(E + 4\pi P) \tag{4.17}$$

The terms H and P can be eliminated by using Eqs. (4.14) and (4.16), and hence we have

$$\frac{k^2 c^2}{\omega^2} E = \left[1 + 4\pi b_{22} + \frac{4\pi b_{12}^2}{(-b_{11} - \omega^2)} \right] E \tag{4.18}$$

and, since E is not equal to zero,

$$\frac{k^2 c^2}{\omega^2} = \varepsilon_\infty + \frac{\varepsilon_0 - \varepsilon_\infty}{\omega_0^2 - \omega^2}$$

Equation (4.18) has two solutions for ω^2, and for both solutions $\mathbf{E}$ is perpendicular to $\mathbf{H}$ and both $\mathbf{E}$ and $\mathbf{H}$ are perpendicular to $\mathbf{k}$ (i.e., transverse). The appropriate dispersion relations (ω versus $\mathbf{k}$) are shown in Fig. 3a.

The various modes are spatially isotropic in our case, but for uniaxial or biaxial crystals this will not be so. However, the theory can be generalized to deal with these lower symmetries.[40,41]

In the absence of free charges, Poynting's theorem gives for the energy flux out of some closed surface σ enclosing a volume τ

$$S = \int (c/4\pi)(\mathbf{E} \times \mathbf{H}) \cdot d\boldsymbol{\sigma}$$

$$= -\int [(4\pi)^{-1}(\mathbf{H} \cdot \dot{\mathbf{H}} + \mathbf{E} \cdot \dot{\mathbf{E}}) + \mathbf{E} \cdot \dot{\mathbf{P}}] \, d\tau \qquad (4.19)$$

Therefore ds/dt, the rate of change of energy density s, is given by

$$\dot{ds/dt} = (4\pi)^{-1}(\mathbf{H} \cdot \dot{\mathbf{H}} + \mathbf{E} \cdot \dot{\mathbf{E}}) + \mathbf{E} \cdot \dot{\mathbf{P}} \qquad (4.20)$$

Thus we have to find for s a form such that Eq. (4.20) holds. Following Huang,[26,27] we try

$$s = \tfrac{1}{2}\dot{\mathbf{W}}^2 - \tfrac{1}{2}b_{11}\mathbf{W}^2 - b_{12}\mathbf{W} \cdot \mathbf{E} - \tfrac{1}{2}b_{22}\mathbf{E}^2 + \mathbf{E} \cdot \mathbf{P} + (8\pi)^{-1}(\mathbf{E}^2 + \mathbf{H}^2) \qquad (4.21)$$

and find that it fulfills this condition since

$$ds/dt = (\dot{\mathbf{W}} \cdot \ddot{\mathbf{W}} - b_{11}\dot{\mathbf{W}} \cdot \mathbf{W} - b_{12}\mathbf{W} \cdot \dot{\mathbf{E}} - b_{12}\dot{\mathbf{W}} \cdot \mathbf{E} - b_{22}\mathbf{E} \cdot \dot{\mathbf{E}}$$

$$+ \dot{\mathbf{E}} \cdot \mathbf{P} + \mathbf{E} \cdot \dot{\mathbf{P}}) + (4\pi)^{-1}(\mathbf{E} \cdot \dot{\mathbf{E}} + \mathbf{H} \cdot \dot{\mathbf{H}}) \qquad (4.22)$$

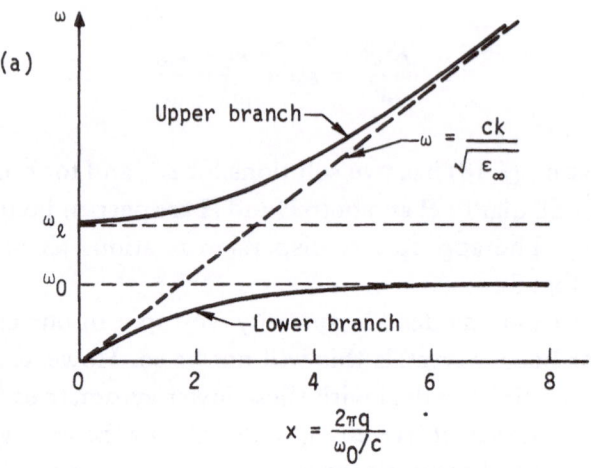

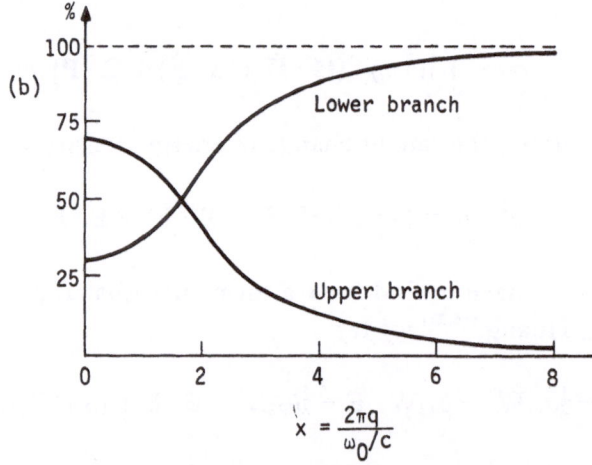

FIG. 3. (a) Schematic frequency versus wave-vector dispersion relations for polaritons in an optically isotropic medium. (b) Partition of energy between mechanical and electromagnetic degrees of freedom for polariton modes.

The last three terms are equal to the expression in Eq. (4.20), and if we group the coefficients of $\dot{\mathbf{W}}$ and $\dot{\mathbf{E}}$, they vanish [cf. Eqs. (4.5) and (4.6)].

Now if we also eliminate $\mathbf{P}$ using Eq. (4.6) and express the b coefficients in terms of macroscopic data, we obtain

$$s = [\tfrac{1}{2}(\dot{\mathbf{W}}^2 + \omega_0^2\mathbf{W}^2)] + [(8\pi)^{-1}(\varepsilon_\infty\mathbf{E}^2 + \mathbf{H}^2)] \qquad (4.21b)$$

where the first term in brackets is "mechanical" energy (i.e., subject to dissipation by lattice anharmonicity) and the second term in brackets is radiative energy, provided ε_∞ is sensibly constant, which implies no electronic absorption or dispersion. The relative magnitudes of these terms, obtained after the substitution of $\mathbf{W}$, $\mathbf{E}$, and $\mathbf{H}$ from the earlier equations, are shown in Fig. 3b.

The foregoing arguments can be repeated if we have free charges present,[39] but we feel that no useful purpose is served by such repetition here.

4.2. Microscopic Theory

This section is designed to relate the b_{ij} coefficients to microscopic parameters and serves as a logical bridge between the long-wavelength limit and the treatment for arbitrary wavelength. We restrict ourselves to cubic structures, including the zinc-blende structure, for simplicity and continuity with subsequent work.

The main problem with constructing the microscopic theory is that of determining the effect of an electric field at a given ion site, including the part of the field that arises from the dipoles in the immediate vicinity. We do not propose to go into this point here in detail since it is adequately discussed by Born and Huang.[42] The basic argument is very simple: the *local* field $\mathbf{E}_{loc}$ is of the form

$$E_{loc\,\alpha} = \sum_{\beta} T_{\alpha\beta} P_{\beta}$$

where

$$\mathbf{P} = \mathbf{p}/v_a$$

$\mathbf{p}$ being the average dipole moment per cell, and P_β excludes the dipole moment due to the ion where the field is being evaluated. For cubic and zinc-blende structures, symmetry arguments and Laplace's equation show that

$$T_{\alpha\beta} = T_{\alpha\alpha}\delta_{\alpha\beta} = (4\pi/3)\delta_{\alpha\beta}$$

Thus

$$\mathbf{E}_{\text{eff}} = \mathbf{E} + (4\pi/3)\mathbf{P}$$

This enables us to write the second macroscopic equation, Eq. (4.6), as follows:

$$\mathbf{P} = v_a^{-1}[Ze(\mathbf{u}_+ - \mathbf{u}_-) + (\alpha_+ + \alpha_-)\mathbf{E}_{\text{eff}}] \qquad (4.22a)$$

where Ze is the "effective" ionic charge, which we shall examine in more detail. We now introduce $\mathbf{W}$ by the equation

$$\mathbf{W} = (\bar{M}/v_a)^{1/2}(\mathbf{u}_+ - \mathbf{u}_-) \qquad (4.22b)$$

where

$$\bar{M} = M_+ M_-/(M_+ + M_-)$$

and using Eqs. (4.22a) and (4.22b), we obtain

$$\mathbf{P} = \left[1 - \frac{4\pi}{3v_a}(\alpha_+ + \alpha_-)\right]^{-1}\left[\frac{Ze}{(\bar{M}v_a)^{1/2}}\mathbf{W} + \frac{\alpha_+ + \alpha_-}{v_a}\mathbf{E}\right]$$

Hence, in Eq. (4.6), we have

$$b_{21} = \frac{Ze(\bar{M}v_a)^{-1/2}}{1 - (4\pi/3)[(\alpha_+ + \alpha_-)/v_a]}$$

$$b_{22} = \frac{(\alpha_+ + \alpha_-)/v_a}{1 - (4\pi/3)[(\alpha_+ + \alpha_-)/v_a]}$$

Equation (4.5) follows similarly. For the motion of the ions, the non-Coulombic restoring force constant is a simple scalar, and thus

$$\mathbf{F}_r = -K(\mathbf{u}_+ - \mathbf{u}_-) \qquad \text{(positive ion)}$$

and

$$\mathbf{F}_r = K(\mathbf{u}_+ - \mathbf{u}_-) \qquad \text{(negative ion)}$$

Therefore, from Newton's second law, we have

$$M_+\ddot{\mathbf{u}}_+ = -K(\mathbf{u}_+ - \mathbf{u}_-) + Ze\mathbf{E}_{\text{eff}}$$

and (4.23)

$$M_-\ddot{\mathbf{u}}_- = K(\mathbf{u}_+ - \mathbf{u}_-) - Ze\mathbf{E}_{\text{eff}}$$

Equations (4.23) give

$$\bar{M}(\ddot{\mathbf{u}}_+ - \ddot{\mathbf{u}}_-) = \left\{ -K + \frac{(4\pi/3)[(Ze)^2/v_a]}{1 - (4\pi/3)[(\alpha_+ + \alpha_-)/v_a]} \right\}(\mathbf{u}_+ - \mathbf{u}_-)$$

$$+ \left\{ \frac{Ze}{1 - (4\pi/3)[(\alpha_+ + \alpha_-)/v_a]} \right\}\mathbf{E}$$

When this is expressed in terms of $\mathbf{W}$, we find that

$$\ddot{\mathbf{W}} = \left\{ \frac{-K}{\bar{M}} + \frac{(4\pi/3)(Ze)^2/(\bar{M}v_a)}{1 - (4\pi/3)[(\alpha_+ + \alpha_-)/v_a]} \right\} \mathbf{W}$$

$$+ \left\{ \frac{Ze/(\bar{M}v_a)^{1/2}}{1 - (4\pi/3)[(\alpha_+ + \alpha_-)/v_a]} \right\} \mathbf{E}$$

[cf. Eq. (4.5)]. Thus

$$b_{11} = \frac{-K}{\bar{M}} + \frac{(4\pi/3)(Ze)^2/(\bar{M}v_a)}{1 - (4\pi/3)[(\alpha_+ + \alpha_-)/v_a]}$$

$$b_{12} = b_{21} = \frac{Ze/(\bar{M}v_a)^{1/2}}{1 - (4\pi/3)[(\alpha_+ + \alpha_-)/v_a]}$$

Note that if $\mathbf{W} \equiv 0$,

$$\varepsilon_\infty = 1 + 4\pi b_{22}$$

and therefore

$$\frac{4\pi}{3v_a}(\alpha_+ + \alpha_-) = \frac{\varepsilon_\infty - 1}{\varepsilon_\infty + 2} \qquad (4.24)$$

which is the Clausius–Mossotti relation.[43]

Now combining the macroscopic and microscopic equations for b_{11} and b_{12}, we obtain

$$\omega_0^2 \left(\frac{\varepsilon_0 - \varepsilon_\infty}{4\pi} \right) = b_{12}^2 = \frac{(Ze)^2/(\bar{M}v_a)}{\{1 - (4\pi/3)[(\alpha_+ + \alpha_-)/v_a]\}^2}$$

and

$$b_{11} = -\frac{K}{\bar{M}} + \left(\frac{4\pi}{3}\omega_0^2\frac{\varepsilon_0 - \varepsilon_\infty}{4\pi}\right)\left(1 - \frac{4\pi}{3}\frac{\alpha_+ + \alpha_-}{v_a}\right)$$

and using Eq. (4.24), we get

$$b_{11} = -\frac{K}{\bar{M}} + \omega_0^2\frac{\varepsilon_0 - \varepsilon_\infty}{\varepsilon_\infty + 2} = -\omega_0^2 \tag{4.25}$$

The compressibility β for a central Coulomb plus nearest-neighbor short-range overlap potential is

$$\frac{1}{\beta} = \left[\frac{r_0}{3}\frac{d}{dr}\left(\frac{r}{3v}\frac{du}{dr}\right)\right]_{r=r_0} = \frac{r_0^2}{9v_a}\left[\frac{-2\alpha_m(Ze)^2}{r_0^3} + M\Phi''(r_0)\right]$$

where $\Phi''(r_0)$ is the second derivative of the short-range potential $\Phi(r)$ evaluated at $r = r_0$, M is the coordination number, u is the lattice energy per cell, α_m is the Madelung constant, v is the unit-cell volume, and r_0 is the equilibrium nearest-neighbor separation.

Then, using the equilibrium condition

$$(Ze)^2(\alpha_m/r_0^2) + M\Phi'(r_0) = 0$$

we find that

$$\frac{1}{\beta} = \frac{Mr_0^2}{9v_a}\left[\Phi''(r_0) + \frac{2\Phi'(r_0)}{r_0}\right]$$

$$= \frac{r_0^2}{3v_a}K \tag{4.26}$$

Hence, using Eq. (4.25), we obtain, setting $b_{11} = -\omega_0^2$,

$$\frac{1}{\beta} = \frac{\bar{M}r_0^2}{3v_a}\frac{\varepsilon_0 + 2}{\varepsilon_\infty + 2}\omega_0^2 \qquad (4.26a)$$

This is known as the first Szigeti relation.[44] Note that b_{12} has been eliminated during this argument and that this result for b_{11} stands on its own.

Agreement between β calculated using Eq. (4.26a) and the experimental value should be regarded with caution: we are comparing dynamic (i.e., anharmonic) data with equations derived from a harmonic theory, whereas we should really be using "mechanical equilibrium" data, as defined by Hardy and Karo in a recent paper;[45] that is, data reduced to the case of a lattice in mechanical equilibrium at 0°K (no zero-point motion). However, the errors are not large for the alkali halides, and we shall ignore them.

So far we have ignored b_{12} by the simple expedient of eliminating it, but we can also form the ratio

$$s = \frac{b_{12\,\text{obs}}}{b_{12\,\text{calc}}}$$

and if b_{12} is calculated with the assumption that $Z = 1$, the deviation from unity is generally very marked ($\sim 25\%$). On the basis of the shell model, this deviation is attributed to the effect of short-range forces, which produce core-shell displacements that reduce the net polarization in an external electric field.[46,47] We, on the other hand, have followed Szigeti and maintain that "shells" and "cores" are a fiction.

Thus we say that the positive ions "punch holes" in the negative ions, and it is the associated "deformation dipoles" that reduce the net polarization. If we place the resultant deformation dipoles at the negative-ion center, the calculated effective charge[48] is reduced by an amount

$$p = \frac{-M}{3}\left[m'(r_0) + \frac{2m(r_0)}{r_0} \right] \qquad (4.27)$$

$m(r)$ being the dipole moment in a given bond.

Inspection of the preceding equations reveals that these are unchanged, except that Ze becomes $(Ze - p)$, which we shall denote by e^*. Thus

$$\frac{(Ze - p)/(\bar{M}v_a)^{1/2}}{1 - (4\pi/3)[(\alpha_+ + \alpha_-)/v_a]} = \left(\frac{\varepsilon_0 - \varepsilon_\infty}{4\pi} \right)^{1/2} \omega_0 \qquad (4.28)$$

and p is thus directly calculable from Eq. (4.28). Table I gives the ratios $\beta_{\text{calc}}/\beta_{\text{obs}}$ and e^*/e calculated from Eqs. (4.26a) and (4.28), respectively.

To determine $m'(r)$ and $m(r)$ in Eq. (4.27) separately, we have to assume some form for $m(r)$. The simplest and most logical choice is

$$m(r) \propto \exp(-r/\rho)$$

where ρ is the Born–Mayer screening distance. This is qualitatively correct, but modifications are possible—modifications that may be significant when r is approximately equal to the ionic radius. Although we have used this form, there is nothing sacred about it.

From the foregoing discussion, it is evident that we neglect intraionic deformation effects; in this we may be right in some cases and wrong in others. Certainly we should find some means of giving the dipoles a more realistic position since, as things stand, we are making a bad multipole expansion. These points will come up for further discussion at later stages of this exposition.

As a conclusion to this section, we consider the classical theory of the absorption of electromagnetic radiation by a lattice whose

TABLE I. Ratios of Compressibilities β Calculated from Eq. (4.26a) to the Observed Values, Together with the Effective Charge Ratios Calculated from Eq. (4.28)

Crystal	Low-temperature model parameters		Room-temperature model parameters	
	β_{calc}/β_{obs}	e^*/e	β_{calc}/β_{obs}	e^*/e
LiF	1.0294	0.8087	0.9930	0.8140
LiCl	1.1492	0.7683	1.0845	0.7679
LiBr	1.4700	0.6602	1.3098	0.7282
LiI	1.4038	0.6779	1.3570	0.7001
NaF	0.9378	0.8285	0.9540	0.8335
NaCl	1.0029	0.7606	1.0534	0.7646
NaBr	1.0669	0.7252	1.0572	0.7329
NaI	1.1440	0.7066	1.1060	0.7341
KF	0.9924	0.8778	0.9792	0.8898
KCl	0.9499	0.7934	0.9553	0.8122
KBr	0.9917	0.7588	0.9608	0.7779
KI	0.8834	0.7243	0.9117	0.7578
RbF	1.0020	0.9199	0.9810	0.9490
RbCl	1.0211	0.8480	0.9309	0.8722
RbBr	0.9492	0.8246	0.8978	0.8398
RbI	0.9796	0.7666	0.9480	0.7790
CsF	1.2302	0.9153	1.1968	0.9463
CsCl	1.0151	0.8333	0.9543	0.8347
CsBr	0.9649	0.8202	0.9274	0.8244
CsI	0.9476	0.7821	0.9151	0.7914

equations of motion are those given earlier [Eqs. (4.5) and (4.6)], with the inclusion of damping, which we shall introduce shortly. If we now consider the expression for the dielectric constant as a function of frequency, we can see from Eq. (4.8a) that the refractive index $n = \varepsilon^{1/2}$ is purely imaginary in the range of frequencies between ω_0 and ω_l. This means that the reflectivity R is 100% throughout this range since

$$R = \left| \frac{n-1}{n+1} \right|^2 \tag{4.28a}$$

This behavior is in qualitative agreement with the observed selective reflectivity of ionic crystals over this range of frequencies. The agreement, however, is not complete since the reflectivity is never exactly unity. It follows that, since the reflectivity is not perfect, some radiation gets into the crystal, where it is strongly absorbed. We have no mechanism to account for this at present. The reason is very simple: no means of dissipating energy from the normal modes of the lattice is built into the theory. This point is also relevant to the discussion of polariton theory, in which there are also nondissipative propagation equations for coupled electrical and mechanical oscillations. It is necessary to allow for some dissipation of energy from the mechanical component of the vibrations in order to obtain any absorption. To treat this problem rigorously requires a detailed quantum-mechanical treatment of lattice anharmonicity such as that carried out by Cowley[49] and by Bilz[50] by the method of double time- and temperature-dependent Green's functions, together with ordinary perturbation theory. This dissipation is an irreversible process, since the energy thus lost ultimately appears as thermal energy. As this is an anharmonic effect, it is strictly outside the terms of reference of the present review. Nonetheless we shall return later to a discussion of certain important aspects of these calculations in connection with the detailed structure of the damping function.

For the present, we merely wish to include the classical dissipative damping by adding to the equation of motion a simple damping term given by

$$- \alpha \frac{d\mathbf{W}}{dt}$$

Thus the equation of motion [Eq. (4.5)] for the displacement parameter $\mathbf{W}$ becomes

$$\ddot{\mathbf{W}} = b_{11}\mathbf{W} - \alpha\dot{\mathbf{W}} + b_{12}\mathbf{E}$$

Now, substituting a time periodic solution $\mathbf{W} = \mathbf{W}_0 \exp(-i\omega t)$, we obtain the following result:

$$-\omega^2 \mathbf{W} = (b_{11} + i\omega\alpha)\mathbf{W} + b_{12}\mathbf{E}$$

and

$$\varepsilon(\omega) = 1 + 4\pi b_{22} + \frac{4\pi b_{12}b_{21}}{(-b_{11} - \omega^2 - i\omega\alpha)}$$

For a plane wave propagating along the positive x direction, we have

$$\mathbf{E} = \mathbf{E}_0 \exp\left(i\omega\left\{[\varepsilon(\omega)]^{1/2}\frac{x}{c} - t\right\}\right) \tag{4.29}$$

We now define a complex refractive index

$$\bar{n} = n(1 + i\kappa) = [\varepsilon(\omega)]^{1/2}$$

The real and imaginary parts of this quantity can be expressed in terms of the material parameters—that is, the dielectric constants and the dispersion frequency—according to the following two equations:

$$n^2(1 - \kappa^2) = \varepsilon_\infty + \frac{(\varepsilon_0 - \varepsilon_\infty)[1 - (\omega/\omega_0)^2]}{[1 - (\omega/\omega_0)^2]^2 + (\omega/\omega_0)^2(\alpha/\omega_0)^2}$$

and

$$2n^2\kappa = \frac{(\varepsilon_0 - \varepsilon_\infty)(\alpha/\omega_0)(\omega/\omega_0)}{[1 - (\omega/\omega_0)^2]^2 + (\omega/\omega_0)^2(\alpha/\omega_0)^2}$$

Thus we now have for the phase factor in Eq. (4.29) the following expression:

$$\exp\left[i\omega\left(\frac{nx}{c} - t\right)\right] \exp\left(-\frac{n\kappa\omega}{c}x\right)$$

and we see that we have a wave propagating in the positive x direction with a phase velocity given by c/n. Moreover, it has an exponential attenuation factor. Therefore we have the possibility of a wave propagating into the medium with attenuation, and this will be very strong for $\omega_0 \lesssim \omega \lesssim \omega_l$. Hence it is now possible to work out the reflectivity according to Eq. (4.28a), with n replaced by $\bar{n} = \varepsilon^{1/2}$. The agreement between theory and experiment is quite reasonable. However, observed reflectivity spectra contain fine structure that is not reproduced by the simple theory. Effectively, this means that a single damping constant is not sufficient. Indeed, one of the ways of interpreting experimental reflectivities is to use a superposition of several dispersion oscillators.

It is not very easy to determine the actual dispersion frequency ω_0 from the reflectivity spectrum. The usual way of determining ω_0 is to examine the transmission of very thin films since, as we shall show, for a sufficiently thin film the minimum in the transmission at normal incidence occurs exactly at ω_0. To examine this problem we must use the full solutions for the Maxwell equations describing the propagation of electromagnetic waves. Thus we will have on the incident side both forward- and backward-propagating waves (electric and magnetic), within the plate a similar pair of attenuated waves, and on the far side only a forward-propagating wave. In all three regions the electric and magnetic fields are related by Eq. (4.16) with k being equal to (ω/c) outside the plate and $\omega\bar{n}/c$ inside.

We now impose the requirement that the electric and magnetic fields be continuous across the two boundaries (this is appropriate when the medium is an insulator) and obtain the following expression for the ratio of the transmitted electric field amplitude E to the incident electric field amplitude A for a plate of thickness d:

$$\frac{E}{A} = \frac{4\bar{n}}{(1+\bar{n})^2 \exp(-i\omega\bar{n}d/c) - (1-\bar{n})^2 \exp(i\omega\bar{n}d/c)}$$

In general, this is a rather cumbersome expression to handle, and a computer fit will be necessary. But when the plate is very thin, we have $2\pi d \ll \lambda = 2\pi c/\omega$. In these circumstances we can expand the exponentials in the denominator, obtaining

$$E/A = \left[1 - \frac{i\omega d}{2c}(1+\bar{n}^2) + \cdots \right]^{-1}$$

Thus the transmission, defined as $T = |E/A|^2$, is given by

$$\left[1 + \frac{i\omega d}{2c}(\bar{n}^{*2} - \bar{n}^2) + \cdots \right]^{-1}$$

and minimum transmission occurs when $i\omega(\bar{n}^{*2} - \bar{n}^2) = i\omega(\varepsilon^* - \varepsilon)$ is maximal.

Substituting the value of ε, we obtain the condition for a maximum:

$$0 = \frac{d}{d\omega}\left[\frac{\omega^2}{(\omega_0^2 - \omega^2)^2 + \alpha^2\omega^2} \right] = \frac{2\omega(\omega_0^2 + \omega^2)(\omega_0^2 - \omega^2)}{[(\omega_0^2 - \omega^2)^2 + \alpha^2\omega^2]^2}$$

This condition is satisfied when $\omega = \omega_0$.

Montgomery and Misho[51] have pointed out that one must be careful of the conditions under which one uses this equation since the normal criteria of thinness are not, in fact, sufficiently stringent to ensure that the expansion in the equation is valid. They have considered the higher-order corrections to this equation and have shown that one can best determine ω_0 from a plot of the frequency of the transmission minimum against the film thickness, extrapolated to the limiting frequency for an infinitely thin film. Values of the dispersion frequencies for all the alkali halides obtained in this way are shown in Table II.

In a later section we will discuss the modifications that are necessary to obtain a detailed insight into the nature of the damping function and the origin of its structure.

TABLE II. Experimental Values of the Infrared Dispersion Frequency $\omega_0{}^a$

Crystal	Infrared dispersion frequency ω_0 (cm^{-1})	
	Low temperature	Room temperature
LiF	318	305
LiCl	221	203
LiBr	187	173
LiI	151.5	142
NaF	262	246.5
NaCl	178	164
NaBr	146	134
NaI	124	116
KF	201.5	194
KCl	151	142
KBr	123	114
KI	109.5	102
RbF	163	158
RbCl	126	116.5
RbBr	94.5	87.5
RbI	81.5	75.5
CsF	134	127
CsCl	106.5	99.5
CsBr	78.5	73.5
CsI	65.5	62.0

[a] Data taken from R. P. Lowndes and D. H. Martin, *Proc. Roy. Soc. (London)* **A308**, 473 (1969); cf. Ref. 134.

5. Description and Justification of the Various Dipolar Models

In Section 3, we discussed the basic theory underlying the Born–Oppenheimer approximation, which allows us to separate nuclear motion from electronic motion. In setting up a model potential function it will often be necessary to include some parameters that specify the ionic polarization and deformation. However, these parameters are always uniquely determined by the instantaneous nuclear configuration, as required by the adiabatic approximation. The situation in metals is somewhat different since the criteria demanded in the preceding section for the validity of the Born–Oppenheimer approximation are not obviously satisfied. Nonetheless, the adiabatic approximation has been used extensively in treatments of the lattice dynamics of metals. In fact, the criteria that we have given previously are excessively stringent, but they are certainly satisfied by the alkali halide crystals. The only novel effect in the dynamics of metals is the existence in the dispersion curves of Kohn anomalies,[52] which have their origin in a sudden change in the effective screening of the interionic potential. This is not a matter that we propose to discuss since it is outside the scope of this review.

From now on we shall take the validity of the adiabatic approximation for granted. Thus, in computing the effective potential function for the nuclear motion, we may evaluate the energy of the system as a function of the nuclear configuration at any given instant and assume that the electronic system thereafter adjusts itself instantaneously to the vibrational motion of the nuclei. Thus the problem of computing the energy of the crystal becomes a quasistatic problem, and this very much simplifies the arguments involved.

We mentioned previously that it is desirable to retain certain electronic degrees of freedom as redundant variables to be determined later. These degrees of freedom are not intended to describe the motion of individual electrons in the crystal. Rather, they are intended to specify the expectation values of certain functions of the electronic coordinates.

There are various ways of discussing this problem. The foundation work on this problem has been carried out by two groups: Tolpygo and co-workers[53-64] and Yamashita and Kurosawa.[65] The two groups have reached similar conclusions, but their approaches are somewhat different. The Soviet group uses perturbation theory to investigate the energy of the crystal as a function of the nuclear coordinates, whereas Yamashita and Kurosawa have used a variational approach. In our opinion, the second type of approach is somewhat clearer and less subject to some of the uncertainties that are introduced by perturbation theory. For example, in Tolpygo's work (see particularly Refs. 53, 58, 60, and 64) the essential idea is to use, as the basis states for the perturbation treatment, the eigenstates of the electronic Hamiltonian for the undisplaced nuclei. For the situation in which the nuclei are displaced, these states are used to compute the energy change by second-order perturbation theory. A rather disturbing feature of this type of treatment, which is probably more apparent than real, is that the excited states of the system are very much more diffuse than the ground states. The excited states are such that one expects the electrons to be largely delocalized. Thus it is not obvious that the perturbation theory converges, and it is difficult from this point of view to satisfy oneself that the dipolar approximation, which we shall discuss in detail shortly, is good.

Yamashita and Kurosawa,[65] using a different approach, introduce variational parameters that are designed to describe the manner in which the electronic state is distorted by the nuclear motion. In this case, no difficulties with convergence are encountered because there is no series expansion, and it can be shown that the higher-order terms that are being neglected are indeed small. Moreover, this approach makes it easier to grasp the physical significance of the various terms in the energy, whereas the corresponding perturbation-theory results are harder to interpret.

This is particularly important from our point of view since it is our hope in this volume to make as clear as possible the various assumptions involved in such calculations. Although a perturbation treatment is, in principle, exact if all the intermediate states are included, in practice the sum over states will have to be truncated. Truncating a perturbation series after only one or two terms gives results that are no better than those obtained by a variational technique involving one or two parameters.

With the exception of the inert-gas solids, the alkali halides represent the nearest approach to the ideal "tight-binding" crystal; that is to say, a crystal whose properties can be discussed in terms of wave functions that are almost completely localized on individual ions. In this situation, in which the functions are localized in direct space, it is better to work with Wannier[66] (i.e., site-orthogonal) functions than with Bloch functions, which are orthogonal in reciprocal space.

In these circumstances the effect of incorporating the ions into the crystal can be regarded as a small perturbation. Correspondingly the effect on the ionic wave functions and energies of any lattice distortion can be regarded as a small localized perturbation that can be discussed in terms of these localized states. If one uses perturbation theory, this point is obscured since the excited states can be very diffuse.

At this point we would like to emphasize that the results of a variational calculation depend on the assumptions that are put in. No new physics is obtained by performing the mathematical manipulation—merely the logical consequences of the initial assumptions. Thus it should be clearly understood that variational calculations are highly constrained by the limited number of variational parameters that can be handled. Since it is likely that the ionic distortions are of a relatively complex character, it follows that one can only provide a relatively crude specification of the actual situation because one has to use a restricted number of variational parameters.

We stress this point because it seems that in the past workers have been somewhat misled by the results of calculations of the variational type (or, for that matter, perturbation-theory calculations) that have appeared to lead to certain effects necessarily being present, whereas these effects originate in the initial assumptions. As an illustration, we refer to the work of Cochran and co-workers[67–69] on the shell model of the alkali halide crystals originally developed by Dick and Overhauser.[46] Originally this was a very mechanistic model of these crystals. Subsequently a number of attempts[70,71] were made to give this model a formal quantum-mechanical justification. As we shall show later, the work of Yamashita and Kurosawa[65] leads to equations very similar to those given by the shell model. On closer examination it is clear that at least some of the calculations that appear to justify the shell model in detail do so because they are equivalent to variational restatements of the basic shell-model assumptions. A more general theory would require more disposable parameters than are allowed by the shell model.

We shall now proceed to discuss both the shell model[46] and our own deformation-dipole model[72–75] on a more or less parallel basis and then describe how these models have some quantum-mechanical justification. The basic assumption both models have in common is that it is adequate to describe the electronic distortion associated with a lattice wave by a point-dipole approximation. Thus it is assumed that the effect of lattice distortion on the electronic configuration can be adequately described by an effective dipole placed at the center of a given ion. This assumption is of central importance to both theories, and its justification is difficult to place on any very firm foundation. For the moment, we shall simply accept this as a good assumption and then proceed to discuss the two models.

The shell model as used by Cochran[67] is a simplified version of the original shell model of Dick and Overhauser.[46] The basic idea is to regard the ion as being composed of a central massive core that contains the nucleus and the inner valence electrons and is attached by isotropic springs to an outer massless spherical shell, taken to represent the valence electrons. Short-range interionic forces are assumed to act between the shells, between the cores, and between the shells and the cores. For the alkali halides, the short-range interionic forces are often restricted to those acting between the shells alone.

It can be seen from this description that the shell model, in principle, introduces a very large number of disposable parameters for each bond in the crystal. Simplifying assumptions have to be made in order to reduce the total number of parameters to manageable proportions (e.g., the restriction that the short-range forces act directly only between shells). This model then leads naturally to the dipolar approximation, since the effect of any lattice distortion on the configuration of a given ion is to displace the core and the shell with respect to each other. The effect of doing this, since the shell is spherical, is to create a dipole at the ion center, to lowest order.

The crucial feature of the shell model is that it will lead to a short-range polarization of a given ion produced by the short-range forces in the crystal. For simplicity, one restricts the short-range forces to act between first neighbors only or, for the alkali halides, between first-neighbor shells. In addition, there is the possibility of shell–core displacements being produced by any effective field that may be present owing to the effective charges of the other ions in the crystal or owing to any other dipoles that are induced on these ions. There are therefore two polarization mechanisms: induction by short-range forces and induction by long-range electric fields. Since this is the case, the model is capable of explaining the deviations of the Szigeti effective charges from the full electronic charge. However, it is a constrained model in that the shells are regarded as rigid. Thus any effective push on one side of the shell is communicated rigidly to the other side. In this it differs from the deformation-dipole model. (Moreover, because of the shell rigidity, there are only three polarization degrees of freedom per ion, and there is thus an inevitable coupling between the field-induced polarizations and those induced by the short-range forces.)

Furthermore, there is a rather more subtle effect, which, though not particularly important for the alkali halides, has proved useful when the shell model has been employed to account for the lattice dynamics of group IV materials. There occurs a non-Coulombic long-range polarization in the crystal because, if one shell is displaced with respect to its core, shell–shell interaction causes this core–shell displacement to be passed on to other atoms in the crystal. This effect has been discussed in detail by Cochran.[76]

More recently, Schröder[77,78] has proposed that one should allow for an additional degree of freedom by allowing for a symmetric breathing distortion of the shells (i.e., the shells are allowed to "breathe" in and out). This modification appears to lead to a significant improvement over the simple shell model. It is introduced in such a way that, though there is an additional degree of freedom due to the breathing motion, there are no additional parameters in the theory. This, of course, is an advantage from the point of view of making calculations of phonon-dispersion curves with minimal numbers of input parameters, but it appears to us that this assumption is somewhat difficult to justify. It amounts to assuming that the shell itself is completely free to breathe; it has no inherent stiffness to oppose its breathing mode. If this is the case, there is no need for an additional parameter; however, this assumption seems implausible since one would expect the existence of a distinct force constant resisting the breathing motion of the shell. (Such additional force constants have been introduced by Sangster and Peckham.[79,80])

The deformation-dipole model is constructed according to a somewhat different philosophy, going back to the original work of Szigeti[44,48] in which he first pointed out the necessity of postulating some type of short-range polarization mechanism to explain the values of the dynamical effective charges for ionic crystals. He made a critical examination of the dielectric properties of the alkali halide crystals, in the course of which he derived two relations: the so-called first and second Szigeti relations, which have been discussed in Section 4 [Eqs. (4.26a) and (4.28)].

The basis of Szigeti's work was to use the dipolar approximation for ionic crystals consistently. When this is done, one is led into a paradox because the dipolar approximation leads logically to the use, for the effective field acting on a given ion, of the relation

$$\mathbf{E}_{\text{eff}} = \mathbf{E} + (4\pi/3)\mathbf{P} \tag{5.1}$$

where $\mathbf{E}_{\text{eff}}$ is the effective field, $\mathbf{E}$ is the macroscopic field, $\mathbf{P}$ is the polarization, and $(4\pi/3)\mathbf{P}$ is the Lorentz local field.

To do this and still assume that the ions carry a full electronic charge $\pm e$ leads to values of the static dielectric constants that are about twice as large as those observed. Szigeti was therefore led to postulate that the ions behave as if they carried an effective charge that was smaller than the full electronic charge. This was somewhat disturbing since the classic work of Born, Huggins, and Mayer[17–21] on the cohesive energies of alkali halide crystals had led one to believe that the cohesion of these materials could be understood by using the classical picture of an array of point charges of magnitude $\pm e$, held apart by short-range repulsions.

Szigeti[48] suggested the probable origin of the discrepancy between the effective charge determined from dielectric-constant data and the full electronic charge. He pointed out that, since there exists a short-range repulsion between first-neighbor ions, it is to be expected that this will affect their electronic configurations. There will be a tendency for electrons on the ions to be extruded from regions of relatively high charge density. In the undistorted crystal, this effect is essentially contained in the short-range part of the cohesive energy and does not really affect the electrostatic energy since the lowest-order multipole associated with this type of distortion is a hexadecapole. Such multipoles interact only over a very short distance. However, the situation is very much changed if we are considering a distorted lattice where the distortion about a given ion site has an antisymmetric component. In this case, a dipole will be induced on the ion (see Fig. 4), and its magnitude can be easily calculated:

$$\mu_{d\alpha} = m'(r_0)[(\xi_1^{\alpha} - \xi_0^{\alpha}) + (\xi_0^{\alpha} - \xi_2^{\alpha})]$$

$$+ \frac{m(r_0)}{r_0} \sum_{\delta=3\to6} (\xi_{\delta}^{\alpha} - \xi_0^{\alpha}) \tag{5.2}$$

This result is appropriate to the rock-salt structure; for the cesium chloride structure the form is different.

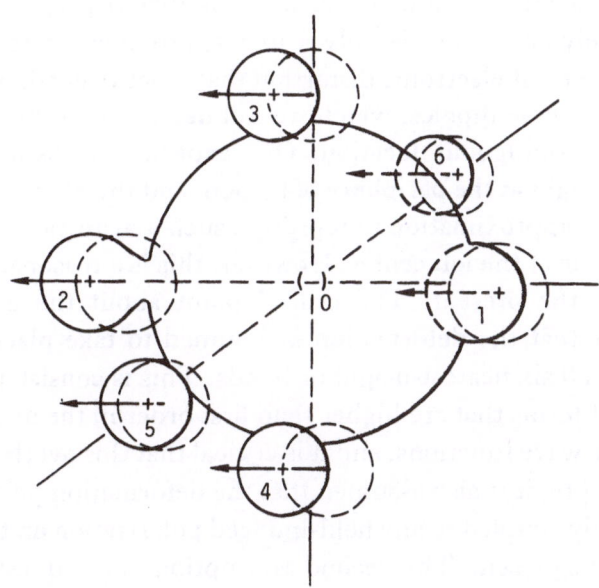

FIG. 4. Schematic representation of the electronic distortion associated with an arbitrary set of ionic displacements for the NaCl lattice.

For a uniform sublattice displacement this dipole couples with the external effective field on the ion and acts in such a way as to change the effective charge by an amount p:

$$\pm e \to \pm (e - p)$$

where

$$p = -\frac{M}{3}\left[m'(r_0) + \frac{2m(r_0)}{r_0} \right]$$

[see Eq. (4.27]. Given Szigeti's hypothesis, one then expects the ions to move in antiphase as if they carried an effective charge smaller than the full electronic charge.

As Szigeti himself makes clear in his 1950 paper,[48] this picture is relatively crude and is only a first approximation to the rather complex type of electronic distortion that is actually taking place. In particular, these dipoles, which we call deformation dipoles, have a somewhat ambiguous location. One expects the distortion to be strongest right at the periphery of the ion, and therefore it is perhaps not a good approximation to represent such a distortion simply by a point dipole at the ion center. However, this is a reasonable starting point for the present. The crucial point about the deformation dipoles is that the deformation is assumed to take place independently in all six nearest-neighbor bonds. This is consistent with the neglect of terms that are higher than first order in the overlap of the electronic wave functions, and it is critical that this overlap be small. Furthermore, it is also assumed that the deformation polarization is not directly coupled to any field-induced polarization on the ion that may also be present. This second assumption can be questioned, but a plausibility argument can be made since one expects the electrons in the region of large overlap, where the deformation polarization is at its largest, to be in a state of local equilibrium under the effect of the short-range forces and therefore to be insensitive, to first order, to the presence of an external electric field.

Thus the deformation-dipole model is to a certain extent more flexible than any of the various shell models or the variational calculations, to be discussed later, that have been used to provide a justification of the shell model. This is so because a full specification of the electronic charge distribution of a given ion in an arbitrarily distorted lattice requires a considerable number of parameters.

Note that the distortion we have just been discussing is restricted to a dipolar type of distortion, but higher multipole types of distortion are also present. Even to specify the dipolar configuration requires several parameters, and these are intimately connected with the way in which the short-range forces exerted by the neighbors on a given ion interact with one another. Throughout this work, we are restricting ourselves to the assumption that the short-range potential is of a pairwise nature. Permitting the dipole in one bond to influence the dipole in another is in principle introducing many-body effects, and this is carrying the theory to a higher level of sophistication than is at present justifiable.

Since the number of parameters is relatively large, it is necessary to make simplifying assumptions in order to reduce their number. The shell model and the deformation-dipole model represent two alternatives that, it is hoped, will represent the two extremes. For example, the shell model says that a push on one side of a given ion is communicated rigidly to the neighbor on the opposite side of the same ion; the deformation-dipole model says that there is no such communication. These are two extreme possibilities; the truth probably lies somewhere in between, and, insofar as the results obtained by the two models are very similar, one can say justifiably that the results are therefore insensitive to these more subtle effects.

Another major question is the validity of the dipolar approximation to the infinite series of odd multipoles that are generated by an antiphase motion of the two sublattices. Again, this is not an easy question to answer, and to our knowledge there is at present no

satisfactory discussion of the problem. Bearing in mind our dis-
tinction between the two types of electronic dipole (i.e., the deformation dipole and the polarization dipole, the latter having its origin
in the presence of a local electric field), it is necessary to discuss the
validity of the dipolar approximation for both types of polarization
separately.

In the case of the polarization dipoles, one has some hope of
justifying the dipolar approximation on the basis of the semiempirical work of Tessman et al.,[81] who showed that it is possible to
derive a set of self-consistent crystal polarizabilities from which the
observed optical-frequency dielectric constants can be derived by
using the Clausius–Mossotti formula,

$$\varepsilon_\infty = 1 + \frac{4\pi[(\alpha_+ + \alpha_-)/v_a]}{1 - (4\pi/3)[(\alpha_+ + \alpha_-)/v_a]} \tag{5.3}$$

where the α terms are host-crystal independent self-consistent
polarizabilities, v_a is the primitive-unit-cell volume, and $\mathbf{P}$ now
contains only the electronic polarization. (Note that the observed
dielectric constants were corrected for dispersion and thus correspond to the dielectric constants at zero frequency with the ions
held fixed.)

That these results were derived by using the Clausius–Mossotti
formula implies that the full Lorentz local-field correction is appropriate. This in turn implies the dipolar approximation. Tessman et
al.[81] also investigated the effect of assuming a different local-field
correction and found that any appreciable deviation from the
$(4\pi/3)\mathbf{P}$ value led to an appreciably less self-consistent set of crystal
polarizabilities.

These results do not of themselves prove the validity of the
dipolar approximation for the field-induced dipoles, but they do
provide a reasonably strong argument in favor of this assumption in
the sense that Tessman et al.[81] have proved that the assumption of
characteristic ionic polarizabilities together with the dipolar approx-

imation does give a consistent set of high-frequency dielectric constants for the alkali halide crystals. It is, of course, possible that there is a fortuitous cancellation of errors and that this result is spurious. However, this is unlikely. More recently Pirenne and Kartheuser[82] have extended the work of Tessman et al.[81] and have shown that, if one allows for a certain coupling between the induced dipole moments on the two types of ion in the crystal, an even better fit to the dielectric constants can be obtained. Again, the full Lorentz local-field correction was assumed. We shall not in our work allow for this correction to the polarizability of the ions, but it can be included in principle. However, recent work by Jaswal and Sharma[83] has indicated that such corrections are probably negligible. (It should be noted that in Jaswal and Sharma's paper the total *relative* mean-square deviation is minimized rather than the *total* mean-square deviation.)

At this point it is worth digressing to refer to the work of Adler[84] (see also Wiser[85]), who has made an examination of the true significance of the local-field correction and has developed a general theory that in principle can be applied in all regimes, depending on whether the electrons are well localized about their ionic sites or are completely delocalized. Adler's theory is based on the use of a generalized Hartree dielectric function, recognizing that a given-wavelength perturbation on the electron distribution can in principle produce a response that has not only the same wave vector but also higher Fourier components with wave vectors given by the applied wave vector plus any vector of the reciprocal lattice. This leads to a dielectric matrix of infinite rank, and it is the off-diagonal terms in this matrix that can be shown to give rise to the local-field correction. However, this theory is very general and at present very far from being in a form that is of practical use.

More recently, Sinha[86] (see also Ref. 71) has given a formal treatment, based on the augmented plane wave (APW) method together with the dielectric matrix, for calculating phonon frequencies. The treatment has been applied to silicon[87,88] with satisfactory results. However, to return to the main theme, it seems to us reasonable in our work to use the dipolar approximation for the field-induced polarization.

There is no corresponding argument for the deformation polarization. Indeed, as we have already seen, the validity of the point-dipole approximation is questionable. In the case of the shell model, as given by Dick and Overhauser,[46] the problem was certainly present, since the model allowed for an additional polarization analogous to the deformation-dipole short-range polarization associated with the overlap between first neighbors. As Dick and Overhauser pointed out, the representation of this overlap polarization, or charge transfer, by point dipoles at the ion centers is extremely suspect.

Thus, in the most general form of the deformation-dipole model, which we have not investigated to date, the location of the deformation dipoles is an open question. The two extremes that are easiest to treat from the point of view of numerical calculation are (1) the extreme where one assumes that the deformation dipoles are located at the center of the negative ion and (2) the extreme where one assumes that they are located at the center of the positive ion.

There is the additional question as to which of the ions deforms or whether in fact both ions deform. We shall discuss this in more detail when we arrive at the presentation of actual results. At this point it can be said fairly conclusively that in every case it is necessary to assume that only the negative ions deform.

As stated above, the effective field experienced by this deformation is still an open question. Thus in some cases it is adequate to place these deformation dipoles on the negative ion at the negative-ion center. However, in certain circumstances there is a reasonable case for locating them at or near the center of the positive ion. This is particularly so in the case of crystals in which the negative ion is very much larger than the positive ion (e.g., sodium iodide), and it is worth noting that this change is rather similar in its effect on the dispersion curves to that produced by Schröder's breathing-shell model.[77,78] It would be possible to divide the deformation dipoles induced on the negative ions between the two types of ion and thus parallel what has been done in some of the later shell-model calculations. The somewhat surprising values for the shell-model parameters that are obtained by the analogous procedure can be readily understood in terms of the deformation-dipole model, for which one obtains the same effect on the dispersion curves by sharing the deformation dipole induced on the negative ion between the positive- and negative-ion centers.

In our calculations we have not considered any intermediate division of the deformation dipole between the two types of ion, but it would certainly be possible to do this. However, one is then in the situation of introducing additional disposable parameters that one is not able to determine without direct reference to the measured dispersion curves.

We turn now to the short-range potential—that is, the Born–Mayer repulsion—which is necessary to maintain the crystal stability. In our first calculation, we followed Kellermann and restricted this interaction to be effective between first neighbors only. In later calculations, we have extended our work to include interactions between second neighbors. In no case does the full potential need to be known. All that is necessary is information about its first and second derivatives at the appropriate neighbor separations. These one can obtain from the equilibrium lattice constant, the observed compressibility, the elastic constant C_{44}, and the observed infrared dispersion frequency ω_0. However, it is not possible to determine from these parameters alone the relative magnitudes of the second-neighbor interactions between positive ions and between negative ions. This is something that has to go into the theory as another disposable parameter, which again can be determined from the measured dispersion curves when these are known. In most of our work, we have restricted the second-neighbor interactions to negative ions only. This is usually a reasonable assumption, since the negative ions are both larger and more polarizable. This last fact leads one to expect the attractive van der Waals interactions between these neighbors to be significantly larger than those between positive ions.

In all this work we assume that the short-range forces are central forces and have made no attempt to include many-body forces, which Löwdin[22] found to exist in his quantum-mechanical treatment of the cohesive energies of ionic crystals and which have been considered by Verma and Singh,[89,90] whose work is an extension of Lundqvist's[91] treatment. Some measure of the importance of these many-body forces is obtained by examining the degree to which the Cauchy relation, $C_{12} = C_{44}$, is or is not satisfied by ionic crystals. The extent to which this relation is satisfied is reasonable for some of the alkali halides, but one cannot say that it is ideally satisfied by any of them.

In order to extend the theory to allow for violations of the Cauchy relation, we have extended the deformation-dipole model to include "angle-bending forces," first introduced by Maradudin[92] in unpublished work (a general derivation is given by Cunningham[93]). This is a force that resists any change of the 90° bond angle between any two first neighbors of a given ion. In principle, there are two constants: one for the sodium–chlorine–sodium angle and the other for the chlorine–sodium–chlorine angle. As a first approximation, we assume the two are the same. This type of force to some extent can be expected to simulate the effects of the many-body forces. It does allow us to fit all three elastic constants, which is an advantage, since it avoids discarding input data that are readily available. The main virtue of using angle-bending forces is that the resultant potential function is rotationally invariant.

An alternative procedure followed by Woods et al.[67] allows the shearing force constant for first-neighbor displacements to be a noncentral type of restoring force; in other words, the equilibrium condition is not imposed on the first derivative of the short-range first-neighbor interaction. To justify this it is necessary to postulate a generalized force that resists only uniform dilation of the lattice. The results of the two procedures are very different. In certain cases the inclusion in the theory of angle-bending forces leads to appreciably better agreement with the observed dispersion curves. (We have neglected any effect that these short-range forces may have on the dipolar interaction.)

Finally, it should be remembered that the deformation polarization as we understand it is associated with only part of the short-range forces: the part that contributes the Born–Mayer overlap repulsion. We do not think the van der Waals component of the short-range interaction will give rise to any appreciable deformation-type polarization.

In order to understand the quantum-mechanical foundation of these various models, we shall now discuss the problem from the point of view of the variational technique used by Yamashita and Kurosawa.[65] This technique has also been employed in a somewhat similar way by Levin and Offenbacher[94] to study the dielectric properties of lithium fluoride.

Initially Yamashita presented a detailed variational calculation of dielectric data and related properties for lithium fluoride.[95] In this calculation the fluoride ions alone were allowed to distort, and the distortion was described by a variational parameter λ. The resultant expression for the energy change ΔE for a uniformly polarized crystal is composed of an internal energy change,

$$\Delta E_1 = a\mathbf{p}_-^2 + b\mathbf{p}_- \cdot \mathbf{p}_\xi + c\mathbf{p}_\xi^2 - \tfrac{1}{2}(4\pi/3v_a)(\mathbf{p}_- + \mathbf{p}_\xi)^2 \quad (5.4)$$

and a term that describes the interaction with the external field:

$$\Delta E_2 = -\mathbf{F} \cdot (\mathbf{p}_- + \mathbf{p}_\xi) \quad (5.5)$$

In these equations, $\mathbf{p}_-$ is the electronic dipole moment of the negative ion, $\mathbf{p}_\xi$ is the dipole moment associated with the relative displacement $\boldsymbol{\xi} = \boldsymbol{\xi}_+ - \boldsymbol{\xi}_-$ of the two sublattices, $\mathbf{F}$ is the external field, v_a is the primitive-unit-cell volume, and a, b, and c are certain material constants.

Equations (5.4) and (5.5) can be used in two ways. In Yamashita's work, the various constants are calculated variationally. This is also the basis of the work by Levin and Offenbacher,[94] who do a similar calculation and correct a certain error they believe to be present in Yamashita's work. This error arises from the fact that in the fluoride ion the distorted $2p$ orbitals are no longer orthogonal to the $1s$ and $2s$ orbitals, and one should correct for this by Schmidt orthogonalization.

The other way in which ΔE, the sum of ΔE_1 and ΔE_2, can be used is as a phenomenological expression in which the various parameters are fitted to experimental data. However, the resultant expression is not sufficiently general to treat all the alkali halides since one wants to include the possibility of positive-ion polarization. Therefore Yamashita and Kurosawa[65] proceeded to give a more general expression,

$$\Delta E = -\mathbf{F} \cdot (\mathbf{p}_+ + \mathbf{p}_- + \mathbf{p}_\varepsilon) - \tfrac{1}{2}(4\pi/3v_a)(\mathbf{p}_+ + \mathbf{p}_- + \mathbf{p}_\varepsilon)^2$$
$$+ a_+\mathbf{p}_+^2 + a_-\mathbf{p}_-^2 + a_{+-}(\mathbf{p}_+ \cdot \mathbf{p}_-) + b_-\mathbf{p}_- \cdot \mathbf{p}_\varepsilon$$
$$+ b_+\mathbf{p}_+ \cdot \mathbf{p}_\varepsilon + c\mathbf{p}_\varepsilon^2 \qquad (5.6)$$

in which the various constants are determined from experimental data. In fact, this expression is now too general in that more constants are present than can be determined uniquely from the empirical dielectric and elastic data.

In the derivation of Eqs. (5.4), (5.5), and (5.6), the dipolar approximation has been used. This in turn requires that the effective field include the Lorentz local field, as can be seen from the form of the second term in Eqs. (5.4) and (5.6). If we now minimize ΔE with respect to the parameters $\mathbf{p}_+$, $\mathbf{p}_-$, and $\mathbf{p}_\varepsilon$, we obtain the following equations:

$$\mathbf{p}_+ = \alpha_+\mathbf{F}_{\text{eff}} \tag{5.7a}$$

$$\mathbf{p}_- = \alpha_-(\mathbf{F}_{\text{eff}} - b_-\mathbf{p}_\varepsilon) \tag{5.7b}$$

$$\mathbf{p}_\varepsilon = (2c)^{-1}(\mathbf{F}_{\text{eff}} - b_-\mathbf{p}_-) \tag{5.7c}$$

and

$$\mathbf{F}_{\text{eff}} = \mathbf{F} + (4\pi/3v_a)(\mathbf{p}_+ + \mathbf{p}_- + \mathbf{p}_\varepsilon) \tag{5.7d}$$

where

$$\alpha_+ = (2a_+)^{-1} \quad \text{and} \quad \alpha_- = (2a_-)^{-1}$$

and we have assumed that a_{+-} and b_+ are zero. The first assumption corresponds to the Tessman–Kahn–Shockley assumption of characteristic polarizabilities; the second corresponds to the neglect of the short-range polarization of the positive ion. One can view these equations in two different ways and obtain either the shell model or the deformation-dipole model. This depends on the significance given to the parameter $\mathbf{p}_\varepsilon$. In general the non-Coulombic force constant K is defined from the change in short-range energy ΔE_{sr} by

$$\Delta E_{\text{sr}} = \tfrac{1}{2}K\xi^2 \tag{5.8}$$

In Yamashita and Kurosawa's expression, $K = 2ce^2$, where $2c$ is given by the relation

$$2c = \frac{\bar{M}}{e^2}\frac{\varepsilon_0 + 2}{\varepsilon_\infty + 2}\omega_0^2$$

$$+ \frac{4\pi}{3v_a}\frac{\varepsilon_\infty + 2}{\varepsilon_\infty - 1}\left(1 - \frac{e^*}{e}\right)^2 \tag{5.9}$$

whereas in Szigeti's work and in our work, K is given by

$$K = \bar{M}\omega_0^2 \frac{\varepsilon_0 + 2}{\varepsilon_\infty + 2} \tag{5.10}$$

In a typical case there is a difference of about 10% between the two values. However, we shall now proceed to show that Yamashita and Kurosawa's work can be interpreted in such a way as to agree with Szigeti's.

The point about Yamashita and Kurosawa's formula given in Eq. (5.6) is that it can be interpreted in two alternative ways. The way in which these authors themselves interpret it can be seen by considering the change in lattice energy that results from compression. They assume that within a uniformly compressed lattice the polarizations $\mathbf{p}_-$, $\mathbf{p}_+$, and $\mathbf{p}_\varepsilon$ are identically zero. This appears to be a valid assumption and leads immediately to the result $K = 2ce^2$ referred to previously.

However, there is an alternative interpretation to the Yamashita–Kurosawa potential, which we shall now discuss. We certainly wish to set the external field $\mathbf{F}$ equal to zero if we are considering a uniform compression, but this does not mean that we have to set the polarizations $\mathbf{p}_+$ and $\mathbf{p}_-$ equal to zero. To see this we will again consider the simplified model in which the parameters b_+ and a_{+-} are set equal to zero. This corresponds to the situation where the negative ion is very much more deformable than the positive ion. Then we have for the energy change the following expression:

$$\Delta E = -\tfrac{1}{2}(4\pi/3v_a)\mathbf{F}_{\text{eff}}^2 + a_+(\alpha_+\mathbf{F}_{\text{eff}})^2 + a_-\alpha_-^2\,(\mathbf{F}_{\text{eff}} - b_-e\boldsymbol{\xi})^2$$

$$+\, b_-\alpha_-(\mathbf{F}_{\text{eff}} - b_-e\boldsymbol{\xi}) \cdot e\boldsymbol{\xi} + ce^2\boldsymbol{\xi}^2 \tag{5.11}$$

where we have used Eqs. (5.7a)–(5.7d) together with the fact that $\mathbf{p}_\varepsilon = e\boldsymbol{\xi}$.

Using the fact that the effective field is equal to zero since $\mathbf{F} = 0$, we finally obtain

$$\Delta E = ce^2\xi^2\left(1 - \frac{b_-^2\alpha_-}{2c}\right) \tag{5.12}$$

and this is equal to

$$\Delta E_{sr} = \tfrac{1}{2}K\xi^2 \tag{5.13}$$

It therefore follows that

$$K = 2ce^2\left(1 - \frac{b_-^2\alpha_-}{2c}\right) \tag{5.14}$$

which differs from the value given by Yamashita and Kurosawa; however, this new value of the constant K is consistent with the value derived by Szigeti.

In the argument just given, we have associated $\mathbf{p}_-$ with the deformation of a given *bond*. Thus the terms that are quadratic in $\mathbf{p}_-$ do not vanish in the case of a uniformly compressed crystal as they would if we regarded $\mathbf{p}_-$ as referring to the polarization of the negative ion as a whole. It should be emphasized that our viewpoint is not the same as that of Yamashita and Kurosawa. The equations are formally the same, but the way in which we have regarded the constituent parameters is different. The importance of this approach is that it does show that one can obtain the equations of the deformation-dipole model from a Hamiltonian that is essentially of the form given by Yamashita and Kurosawa.

From now on we shall confine ourselves to the deformation-dipole model. In a more recent paper by Hardy and Lidiard[96] the original formalism of Hardy[72,73] has been somewhat revised in that an actual deformation-dipole potential function is given explicitly.

The earlier papers were simply concerned with including the deformation-dipole effects in the equations of motion for the harmonic lattice. As we shall see, the deformation-dipole equations are a generalization, designed to cope with lattice waves of arbitrary wavelength, of the ideas developed by Szigeti and Huang for long waves.

Thus we shall now develop our equations of motion using a potential function. The potential function that gives the deformation-dipole equations of motion is

$$\Delta E = \tfrac{1}{2}\tilde{\boldsymbol{\xi}}(\mathbf{R} - \mathbf{H})\boldsymbol{\xi} - \tfrac{1}{2}\tilde{\boldsymbol{\mu}}'\mathbf{U}\mathbf{H}\boldsymbol{\xi} - \tfrac{1}{2}\tilde{\boldsymbol{\xi}}\mathbf{H}\mathbf{U}\boldsymbol{\mu}' - \tfrac{1}{2}\tilde{\boldsymbol{\mu}}'\mathbf{U}\mathbf{H}\mathbf{U}\boldsymbol{\mu}' + \tfrac{1}{2}\tilde{\boldsymbol{\mu}}\mathbf{a}^{-1}\boldsymbol{\mu}$$

$$(5.15)$$

where

$$\boldsymbol{\mu}' = \boldsymbol{\mu} + \boldsymbol{\mu}_d = \boldsymbol{\mu} + \check{\mathbf{S}}\boldsymbol{\xi}$$

Here the $\boldsymbol{\mu}$ and $\boldsymbol{\xi}$ terms are $6N$-component column vectors, whereas the $\mathbf{H}$, $\mathbf{U}$, and $\mathbf{a}$ terms are $6N \times 6N$ matrices if we are dealing with a lattice containing N primitive cells. The $\boldsymbol{\xi}$ terms are the ionic displacements, the $\boldsymbol{\mu}$ terms are their field-induced dipole moments, and the $\boldsymbol{\mu}_d$ terms are the deformation-dipole moments. Furthermore,

$$\mathbf{a} = \alpha_\lambda \delta_{\kappa\lambda}$$

and

$$\mathbf{U} = e_\lambda^{-1} \delta_{\kappa\lambda}$$

where α_λ is the polarizability of the ion λ and e_λ is the corresponding monopole charge [$\lambda \equiv \binom{l}{k}_\alpha$ in the usual notation]; $\mathbf{R}$ and $\mathbf{H}$ are the matrices of short-range and dipole–dipole interactions, respectively.

The various terms in Eq. (5.15) can be simply understood as short-range interactions and the various dipole–dipole interactions, while the final term is the self-energy of the polarization dipoles. The crucial feature of the potential function ΔE is that it includes the deformation polarization, which is described by the matrix $\mathbf{S}$ and its transpose. This potential function can be regarded as a modification of that obtained when one only considers field-induced dipoles; that is, the polarization-dipole model,[72,75]

$$\Delta E = \tfrac{1}{2}\tilde{\boldsymbol{\xi}}(\mathbf{R} - \mathbf{H})\boldsymbol{\xi} - \tfrac{1}{2}\tilde{\boldsymbol{\mu}}\mathbf{U}\mathbf{H}\boldsymbol{\xi} - \tfrac{1}{2}\tilde{\boldsymbol{\xi}}\mathbf{H}\mathbf{U}\boldsymbol{\mu} - \tfrac{1}{2}\tilde{\boldsymbol{\mu}}\mathbf{U}\mathbf{H}\mathbf{U}\boldsymbol{\mu} + \tfrac{1}{2}\tilde{\boldsymbol{\mu}}\,\boldsymbol{\alpha}^{-1}\boldsymbol{\mu} \quad (5.16)$$

The essential modification is that on the right-hand side of Eq. (5.15), $\boldsymbol{\mu}$ is replaced by $\boldsymbol{\mu}'$, defined by

$$\boldsymbol{\mu}' = \boldsymbol{\mu} + \check{\mathbf{S}}\boldsymbol{\xi} \quad (5.17)$$

except in the self-energy term, where one continues to retain only the field-induced dipoles $\boldsymbol{\mu}$. (The justification for this is that the deformation-dipole self-energy is already implicitly contained in the repulsive interaction energy.)

The equations given above are written in terms of direct space quantities using a matrix notation. If we follow the adiabatic principle, we can now write the equations of motion:

$$-\frac{\partial \Delta E}{\partial \xi_\lambda} = m_\lambda \ddot{\xi}_\lambda$$

and

$$-\frac{\partial \Delta E}{\partial \mu_\lambda} = 0$$

Thus

$$[\mathbf{R} - (1 + \mathbf{SU})\mathbf{H}(1 + \mathbf{U\check{S}})]\boldsymbol{\xi} - (1 + \mathbf{SU})\mathbf{HU}\boldsymbol{\mu} = -\mathbf{m}\ddot{\boldsymbol{\xi}} \quad (5.18)$$

where

$$\mathbf{m} = m_\lambda \delta_{\kappa\lambda}$$

and

$$(\mathbf{a}^{-1} - \mathbf{UHU})\boldsymbol{\mu} - \mathbf{UH}(1 + \mathbf{U\check{S}})\boldsymbol{\xi} = 0 \qquad (5.19)$$

It will be observed that at this stage we still have the field-induced dipole moments present as additional degrees of freedom. However, the constraint embodied in Eq. (5.19), which follows from the adiabatic approximation, allows the elimination of these dipoles from further consideration. On doing this, we obtain

$$[\mathbf{R} - (1 + \mathbf{SU})\mathbf{HUC}^{-1}(1 + \mathbf{U\check{S}})]\boldsymbol{\xi} = -\mathbf{m}\ddot{\boldsymbol{\xi}} = \mathbf{m}\omega^2\boldsymbol{\xi} \quad (5.20)$$

where

$$\mathbf{C} = 1 - \mathbf{aUHU}$$

and we have assumed that $\boldsymbol{\xi}$ has a sinusoidal time dependence, as is normal (see Section 2).

These are the final equations of motion in direct space; they are subjected to a Fourier transformation, after which they have exactly the same form except that the matrices, instead of forming one full $6N \times 6N$ matrix, are formed of N 6×6 blocks, one for each allowed wave vector. The explicit form of the Fourier-transformed matrix $\mathbf{H}$ will be discussed later; at this point we only give the matrices for the transforms of $\mathbf{R}$, $\mathbf{S}$, $\mathbf{U}$, and $\mathbf{a}$.

The matrix **R** is made up of three components: (1) that due to nearest-neighbor forces, (2) that due to second-neighbor central forces, and (3) that induced by forces that resist changes in the 90° angles between first-neighbor bonds. Thus

$$\mathbf{R} = \mathbf{m}^{1/2}(\mathbf{R}^{NN} + \mathbf{R}^{NNN} + \mathbf{R}^{NC})\mathbf{m}^{1/2}$$

where

$$\left[R\binom{11}{xx} \right]^{NN} = \frac{e^2}{m_1 v_a}(A' + 2BC')$$

$$\left[R\binom{11}{yy} \right]^{NN} = \frac{e^2}{m_1 v_a}(A' + 2BC')$$

$$\left[R\binom{11}{zz} \right]^{NN} = \frac{e^2}{m_1 v_a}(A' + 2BC')$$

and

$$\left[R\binom{22}{xx} \right]^{NN} = \frac{e^2}{m_2 v_a}(A' + 2BC')$$

$$\left[R\binom{22}{yy} \right]^{NN} = \frac{e^2}{m_2 v_a}(A' + 2BC')$$

$$\left[R\binom{22}{zz} \right]^{NN} = \frac{e^2}{m_2 v_a}(A' + 2BC')$$

Furthermore,

$$\left[R\!\begin{pmatrix}12\\xx\end{pmatrix}\right]^{NN} = -\frac{e^2}{(m_1 m_2)^{1/2} v_a}[A' \cos 2\pi q_x r_0$$

$$+ BC'(\cos 2\pi q_y r_0 + \cos 2\pi q_z r_0)]$$

$$\left[R\!\begin{pmatrix}12\\yy\end{pmatrix}\right]^{NN} = -\frac{e^2}{(m_1 m_2)^{1/2} v_a}[A' \cos 2\pi q_y r_0$$

$$+ BC'(\cos 2\pi q_z r_0 + \cos 2\pi q_x r_0)]$$

$$\left[R\!\begin{pmatrix}12\\zz\end{pmatrix}\right]^{NN} = -\frac{e^2}{(m_1 m_2)^{1/2} v_a}[A' \cos 2\pi q_z r_0$$

$$+ BC'(\cos 2\pi q_x r_0 + \cos 2\pi q_y r_0)]$$

The $\mathbf{R}^{NNN}$ matrix is real and symmetric and is defined as follows:

$$\mathbf{R}^{```} = \begin{bmatrix} \lambda\frac{m_2}{m_1}\left[R\!\begin{pmatrix}22\\xx\end{pmatrix}\right]^{```} & \lambda\frac{m_2}{m_1}\left[R\!\begin{pmatrix}22\\xy\end{pmatrix}\right]^{```} & \lambda\frac{m_2}{m_1}\left[R\!\begin{pmatrix}22\\xz\end{pmatrix}\right]^{```} & 0 & 0 & 0 \\ & \lambda\frac{m_2}{m_1}\left[R\!\begin{pmatrix}22\\yy\end{pmatrix}\right]^{```} & \lambda\frac{m_2}{m_1}\left[R\!\begin{pmatrix}22\\yz\end{pmatrix}\right]^{```} & 0 & 0 & 0 \\ & & \lambda\frac{m_2}{m_1}\left[R\!\begin{pmatrix}22\\zz\end{pmatrix}\right]^{```} & 0 & 0 & 0 \\ & & & (1-\lambda)\left[R\!\begin{pmatrix}22\\xx\end{pmatrix}\right]^{```} & (1-\lambda)\left[R\!\begin{pmatrix}22\\xy\end{pmatrix}\right]^{NNN} & (1-\lambda)\left[R\!\begin{pmatrix}22\\xz\end{pmatrix}\right]^{NNN} \\ & & & & (1-\lambda)\left[R\!\begin{pmatrix}22\\yy\end{pmatrix}\right]^{```} & (1-\lambda)\left[R\!\begin{pmatrix}22\\yz\end{pmatrix}\right]^{NNN} \\ & & & & & (1-\lambda)\left[R\!\begin{pmatrix}22\\zz\end{pmatrix}\right]^{NNN} \end{bmatrix}$$

where

$$\left[R\!\begin{pmatrix}22\\xx\end{pmatrix}\right]^{NNN}$$

$$= \frac{e^2}{m_2 v_a}[2A'' + 4B'' - A'' \cos 2\pi q_x r_0(\cos 2\pi q_y r_0 + \cos 2\pi q_z r_0)$$

$$- B''(2 \cos 2\pi q_y r_0 \cos 2\pi q_z r_0 + \cos 2\pi q_x r_0 \cos 2\pi q_y r_0$$

$$+ \cos 2\pi q_x r_0 \cos 2\pi q_z r_0)]$$

$$\left[R\binom{22}{yy}\right]^{NNN}$$

$$= \frac{e^2}{m_2 v_a}[2A'' + 4B'' - A'' \cos 2\pi q_y r_0 (\cos 2\pi q_z r_0 + \cos 2\pi q_x r_0)$$

$$- B''(2\cos 2\pi q_z r_0 \cos 2\pi q_x r_0 + \cos 2\pi q_y r_0 \cos 2\pi q_z r_0$$

$$+ \cos 2\pi q_y r_0 \cos 2\pi q_x r_0)]$$

$$\left[R\binom{22}{zz}\right]^{NNN}$$

$$= \frac{e^2}{m_2 v_a}[2A'' + 4B'' - A'' \cos 2\pi q_z r_0 (\cos 2\pi q_x r_0 + \cos 2\pi q_y r_0)$$

$$- B''(2\cos 2\pi q_x r_0 \cos 2\pi q_y r_0 + \cos 2\pi q_z r_0 \cos 2\pi q_x r_0$$

$$+ \cos 2\pi q_z r_0 \cos 2\pi q_y r_0)]$$

$$\left[R\binom{22}{xy}\right]^{NNN} = \frac{e^2}{m_2 v_a}[(A'' - B'') \sin 2\pi q_x r_0 \sin 2\pi q_y r_0]$$

$$\left[R\binom{22}{xz}\right]^{NNN} = \frac{e^2}{m_2 v_a}[(A'' - B'') \sin 2\pi q_x r_0 \sin 2\pi q_z r_0]$$

$$\left[R\binom{22}{yz}\right]^{NNN} = \frac{e^2}{m_2 v_a}[(A'' - B'') \sin 2\pi q_y r_0 \sin 2\pi q_z r_0]$$

In the $\mathbf{R}^{NNN}$ matrix, the coefficient λ is used to adjust the relative magnitudes of the positive–positive and negative–negative contributions.

Finally, the elements of $\mathbf{R}^{NC}$ are given by

$$\left[R\binom{11}{xy}\right]^{NC} = \frac{e^2}{m_1 v_a}(2C \sin 2\pi q_x r_0 \sin 2\pi q_y r_0)$$

$$\left[R\binom{11}{xz}\right]^{NC} = \frac{e^2}{m_1 v_a}(2C \sin 2\pi q_x r_0 \sin 2\pi q_z r_0)$$

$$\left[R\binom{11}{yz}\right]^{NC} = \frac{e^2}{m_1 v_a}(2C \sin 2\pi q_y r_0 \sin 2\pi q_z r_0)$$

$$\left[R\binom{22}{xy}\right]^{NC} = \frac{e^2}{m_2 v_a}(2C \sin 2\pi q_x r_0 \sin 2\pi q_y r_0)$$

$$\left[R\binom{22}{xz}\right]^{NC} = \frac{e^2}{m_2 v_a}(2C \sin 2\pi q_x r_0 \sin 2\pi q_z r_0)$$

$$\left[R\binom{22}{yz}\right]^{NC} = \frac{e^2}{m_2 v_a}(2C \sin 2\pi q_y r_0 \sin 2\pi q_z r_0)$$

$$\left[R\binom{11}{xx}\right]^{NC} = \left[R\binom{22}{xx}\right]^{NC} = 0$$

$$\left[R\binom{11}{yy}\right]^{NC} = \left[R\binom{22}{yy}\right]^{NC} = 0$$

$$\left[R\binom{11}{zz}\right]^{NC} = \left[R\binom{22}{zz}\right]^{NC} = 0$$

The relationships of the coefficients A', B', etc., to short-range potential derivatives and to the experimental data are described in detail later [see Eqs. (7.12)–(7.17)]. We have purposely expressed the **R** matrix in terms of the contributions of the various short-range forces to the *dynamical matrix* [cf. Eq. (2.28)]. This was done because it is the latter that are actually used in computing the normal mode frequencies.

The **S** matrix is defined as follows:

$$[\mathbf{S}] \equiv -b(k)\delta_{\alpha\beta}\delta_{kk'} + a_\alpha(k')\delta_{\alpha\beta}(1-\delta_{kk'})$$

$$= \begin{bmatrix} -b(1) & 0 & 0 & a_x(2) & 0 & 0 \\ 0 & -b(1) & 0 & 0 & a_y(2) & 0 \\ 0 & 0 & -b(1) & 0 & 0 & a_z(2) \\ a_x(1) & 0 & 0 & -b(2) & 0 & 0 \\ 0 & a_y(1) & 0 & 0 & -b(2) & 0 \\ 0 & 0 & a_z(1) & 0 & 0 & -b(2) \end{bmatrix}$$

where

$$a_x(1) = 2[\gamma_1' \cos 2\pi q_x r_0 + \gamma_1(\cos 2\pi q_y r_0 + \cos 2\pi q_z r_0)]$$

$$a_y(1) = 2[\gamma_1' \cos 2\pi q_y r_0 + \gamma_1(\cos 2\pi q_z r_0 + \cos 2\pi q_x r_0)]$$

$$a_z(1) = 2[\gamma_1' \cos 2\pi q_z r_0 + \gamma_1(\cos 2\pi q_x r_0 + \cos 2\pi q_y r_0)]$$

$$a_x(2) = 2[\gamma_2' \cos 2\pi q_x r_0 + \gamma_2(\cos 2\pi q_y r_0 + \cos 2\pi q_z r_0)]$$

$$a_y(2) = 2[\gamma_2' \cos 2\pi q_y r_0 + \gamma_2(\cos 2\pi q_z r_0 + \cos 2\pi q_x r_0)]$$

$$a_z(2) = 2[\gamma_2' \cos 2\pi q_z r_0 + \gamma_2(\cos 2\pi q_x r_0 + \cos 2\pi q_y r_0)]$$

and

$$b(1) = 2(\gamma_1' + 2\gamma_1)$$

$$b(2) = 2(\gamma_2' + 2\gamma_2)$$

Here

$$\gamma_k = m_k(r_0)/r_0$$

and

$$\gamma'_k = \left(\frac{dm_k(r)}{dr}\right)_{r=r_0}$$

Finally,

$$\mathbf{U} = (e_k)^{-1}\delta_{kk'}\delta_{\alpha\beta}$$

and

$$\mathbf{a} = \alpha_k\delta_{kk'}\delta_{\alpha\beta}$$

where α_k is the crystal polarizability of the kth ion.

The form of the **S** matrix has been derived by Fourier transformation of Eq. (5.2). This is the most general form of the **S** matrix, allowing for the deformation of both positive and negative ions. In subsequent work, we shall be using one of the more specialized forms in which the deformation of only the negative ion is considered. However, by appropriate manipulation of the **S** matrix, it is possible to change the location of the deformation dipoles from the center of the negative ion to the center of the positive ion.

At this stage it is convenient to consider the corresponding shell-model equations of motion as given by Woods et al.[67]:

$$m\omega^2\boldsymbol{\xi} = (\bar{\mathbf{R}} + \mathbf{Z}\bar{\mathbf{C}}\mathbf{Z})\boldsymbol{\xi} + (\mathbf{T} + \mathbf{Z}\bar{\mathbf{C}}\mathbf{Y})\mathbf{W}$$

$$0 = (\tilde{\mathbf{T}} + \mathbf{Y}\bar{\mathbf{C}}\mathbf{Z})\boldsymbol{\xi} + (\bar{\mathbf{S}} + \mathbf{Y}\bar{\mathbf{C}}\mathbf{Y})\mathbf{W}$$

where $\bar{\mathbf{C}} = -\mathbf{U}\mathbf{H}\mathbf{U}$, $\mathbf{Y}\mathbf{W}$ is the electronic dipole moment on the ions, $\mathbf{Z} = \mathbf{U}^{-1}$, and $\bar{\mathbf{R}}$, $\mathbf{T}$, and $\bar{\mathbf{S}}$ are matrices of short-range forces.

In this case there are six degrees of freedom per ion: three displacement and three dipolar. It has been shown by Cowley *et al.*[69] that the two force constant matrices differ by a term

$$\tfrac{1}{2}\mathbf{T}\bar{\mathbf{S}}^{-1}\tilde{\mathbf{T}}$$

which represents the effects we have discussed earlier (i.e., such things as the rigid communication of a thrust on one side of the ion to the opposite side). Also there is a coupling between the dipolar displacement and the short-range interaction. (However, as Szigeti[48] has pointed out, if one allows for the overlap repulsion, it is necessary that there be some deformation of the charge clouds between the ions. This is not truly represented by the shell model, since the entire dipolar self-energy appears in the shell-model Hamiltonian.)

This effect was recognized by Dick and Overhauser[46] in their original treatment when they allowed for distortion polarization. However, they encountered the same difficulty we have: the location of the deformation polarization is uncertain, and it is obvious from their work that the dipolar approximation to the deformation polarization is very suspect. Moreover, their work leads them to a picture in which the overlap produces a localized distortion at some point in the bond between first neighbors. This distortion is equivalent to a small local positive charge, which is compensated for by an overall uniform increase in the magnitude of the negative charge on the shells of both positive and negative ions. Within the dipolar approximation, the effective deformation dipole thus produced is the difference between opposed dipoles centered on the two different types of ion.

We wish to stress at this point that there is an inherent difference between our deformation dipoles and the exchange-charge polarization in the Dick–Overhauser model. Our dipoles are to be regarded as arising from *internal* distortions of the negative ion. These have the effect of producing a dipole moment in each bond, and this dipole moment is confined to the negative ion. We thus avoid the difficulty that arises when one has a net deformation dipole that is the difference between two opposed dipoles. This last possibility was discussed in an attempt[97] to develop a heuristic theory capable of assessing the relative magnitudes of the dipoles on the positive and negative ions. When this idea was tested by computing dispersion curves,[73] it was found that any attempt to put a deformation dipole on the positive ion directed toward the negative ion resulted in a marked deterioration in the calculated frequency spectra and the dispersion curves as compared with experimentally derived quantities.

In principle, the deformation-dipole model does introduce terms of a many-body nature into the lattice potential energy. That is, the interaction between two nearest-neighbor ions, because it has an associated dipole moment in the bond, will be influenced by the configuration of the other ions in the lattice since this dipole will interact with the electrostatic potential changes that the displacements of the other ions produce. To the lowest approximation, this is avoided by placing the dipoles at the center of one or the other type of ion. There is then no problem because the many-body effects are represented by the deformation-dipole interaction with the other charges in the lattice.

If this simplification is not made, additional problems arise. The magnitudes of these dipoles will change in a symmetric or breathing type of motion (i.e., a uniform compression of the lattice). Moreover, they will change in shear deformation, giving rise to quadrupole interactions and, more generally, to even-multipole interactions, which will affect the elastic constants and lead to violations of the Cauchy relations. This type of effect was originally considered by Herpin,[98] who found that the resultant deviations from the Cauchy relations were of the wrong sign. Subsequently a similar calculation was made by Lothe,[99] who allowed for both the quadrupolar effects and also for the effects of Löwdin's many-body forces. In this way, Lothe was able to proceed some way toward explaining the observed violations of the Cauchy relations. Somewhat later Dick[100] made an investigation of the exchange-charge effects on the elastic constants and found qualitative agreement between theory and experiment. However, it can be shown that, within the exchange-charge formalism, it is not possible to explain both the magnitudes of the Szigeti effective charges and the deviations from the Cauchy relations: if one adjusts the exchange charges to fit the observed Szigeti charges, the deviations from the Cauchy relations become much too large.

We have presented in this section a survey of the various dipolar models. Their advantage is their essential simplicity. What is needed is a model that will represent the dynamics of the alkali halide crystals with reasonable precision without the necessity of including so many disposable parameters that the model is completely useless for crystals whose dispersion curves have not been measured.

Moreover, it should be remembered that, in the case of models that are fitted to the dispersion curves, the dispersion curves are usually measured only for restricted classes of directions, usually the ⟨100⟩, ⟨110⟩, and ⟨111⟩ directions, and it is not necessarily true that a model that reproduces the dispersion curves along these directions is going to be equally good throughout the zone. In calculations by Benedek and Maradudin[101] on defect vibrations, it appeared that a deformation-dipole model that had been adjusted to fit certain of the critical-point phonon frequencies reproduced the measured infrared-absorption spectrum associated with the chloride ion in potassium iodide with more precision than an extended shell model, least-squares fitted to a restricted set of measured dispersion curves.

6. Derivation of the Dipolar Coupling Coefficients

We have in our equations the matrix $\mathbf{H}$, which represents the Fourier transform of the dipole–dipole interactions between the ions in the lattice. This matrix alone is necessary in order to describe the long-range coupling since the other types of dipolar interaction can be readily incorporated into the dynamical matrix [see Section 5, Eqs. (5.15) and (5.20)].

It is therefore of great importance to be able to evaluate the elements of the **H** matrix. This is a nontrivial problem since the sums in the elements of this matrix have to be carried out over all the ions in the lattice. This complicates matters considerably, as the sums are only conditionally convergent. The most general technique for performing this type of sum is that due to Ewald[24,25,102,103] and subsequently adopted by Kellermann[23] for the evaluation of the elements of the **H** matrix. These are also known as the Kellermann or Coulomb coupling coefficients, denoted by

$$
{}^{c}\begin{bmatrix} kk' \\ xy \end{bmatrix} = H\begin{pmatrix} kk' \\ xy \end{pmatrix}
$$

for each wave vector **q**.

We shall now proceed to describe Kellermann's work in detail. The basis of the technique is the transformation of the single sum over the lattice sites in direct space into two rapidly convergent sums, one over direct space and the other over reciprocal space. One can regard this as a mathematical technique, the theta-function transformation, or one can present a more physical picture of the situation, as follows: in addition to the real point charge, at each lattice site are placed two fictitious charge distributions, one positive and one negative. Both of these are three-dimensional Gaussians and have the same half-width. Then we group the point charge with the Gaussian of opposite sign and sum the fields at a given point due to these entities over direct space. Because of the charge neutrality of each entity, the associated fields are of relatively short range. For the residual Gaussian charge distributions, we sum the total charge distribution over the whole crystal and take its Fourier transform. Thence, using Poisson's equation, we derive the corresponding Fourier components of the potential. From this we can easily compute the field at any point, since the Fourier series is rapidly convergent in reciprocal space.

It is best that we present the derivation as a mathematical technique, and we must observe that this technique is restricted to a lattice whose total charge is zero; otherwise the zero-wave-vector term in the sum over reciprocal space diverges.

For the Coulomb potential, we have

$$
{}^{c}\!\begin{bmatrix} kk' \\ xy \end{bmatrix} = e_{k'}e_{k} \lim_{\mathbf{r}\to -\mathbf{r}_{h'k}} \sum_{l} \frac{\partial^2}{\partial x \, \partial y} f(\mathbf{r}-\mathbf{a}') \exp\left\{ 2\pi i \left[\mathbf{q} \cdot \mathbf{r}\left(\begin{smallmatrix} l \\ k'k \end{smallmatrix}\right) \right] \right\}
$$

$$
{}^{c}\!\begin{bmatrix} kk \\ xy \end{bmatrix} = e_{k}^{2} \lim_{\mathbf{r}\to 0} \left\{ \sum_{l} \frac{\partial^2}{\partial x \, \partial y} f(\mathbf{r}-\mathbf{a}') \exp\left[2\pi i \left(\mathbf{q} \cdot \mathbf{a}' \right) \right] - \frac{\partial^2}{\partial x \, \partial y} f(\mathbf{r}) \right\}
$$

where

$$
f(\mathbf{r}) = |\mathbf{r}|^{-1} \quad \text{and} \quad \mathbf{r}_{kk'} = \mathbf{r}\left(\begin{smallmatrix} 0 \\ k \end{smallmatrix}\right) - \mathbf{r}\left(\begin{smallmatrix} 0 \\ k' \end{smallmatrix}\right)
$$

We define functions

$$
F^{\mathbf{q}}(\mathbf{r}) = \sum_{l} f(\mathbf{r}-\mathbf{a}') \exp[2\pi i (\mathbf{q} \cdot \mathbf{a}')]
$$

and

$$
F_{xy}^{\mathbf{q}}(\mathbf{r}) = \frac{\partial^2}{\partial x \, \partial y} F^{\mathbf{q}}(\mathbf{r}) = \sum_{l} \frac{\partial^2}{\partial x \, \partial y} f(\mathbf{r}-\mathbf{a}') \exp[2\pi i (\mathbf{q} \cdot \mathbf{a}')]
$$

Now

$$
\lim_{\mathbf{r}\to -\mathbf{r}_{h'k}} = \lim_{\mathbf{r}\to \mathbf{r}_{kk'}}
$$

since

$$- \mathbf{r}_{k'k} = \mathbf{r}_{kk'}$$

and

$$\left| \mathbf{r} \binom{l}{k'k} \right| = \left| -\mathbf{r} \binom{l}{kk'} \right| = \left| -\mathbf{r}_{k'k} - \mathbf{a}^l \right| = \left| \mathbf{r}_{kk'} - \mathbf{a}^l \right| = \left| \mathbf{a}^l - \mathbf{r}_{kk'} \right|$$

and

$$\left| \mathbf{r} \binom{l}{k'k} \right| = \left| -\mathbf{r}_{k'k} - \mathbf{a}^l \right| = \left| \mathbf{r}_{k'k} + \mathbf{a}^l \right|$$

Therefore

$$\left| \mathbf{r} \binom{l}{kk'} \right| = \left| \mathbf{a}^l - \mathbf{r}_{kk'} \right|$$

and hence

$$^c \begin{bmatrix} kk' \\ xy \end{bmatrix} = e_{k'} e_k \lim_{\mathbf{r} \to -\mathbf{r}_{k'k}} \sum_l \frac{\partial^2}{\partial x \, \partial y} f(\mathbf{r} - \mathbf{a}^l)$$

$$\times \exp \left\{ 2\pi i \left[\mathbf{q} \cdot \mathbf{r} \binom{l}{k'k} \right] \right\} \qquad (k \neq k')$$

$$= e_{k'} e_k \lim_{\mathbf{r} \to \mathbf{r}_{kk'}} \sum_l \frac{\partial^2}{\partial x \, \partial y} f(\mathbf{r} - \mathbf{a}^l)$$

$$\times \exp \{ 2\pi i \left[\mathbf{q} \cdot (\mathbf{a}^l - \mathbf{r}_{kk'}) \right] \}$$

$$= e_{k'} e_k \sum_l \left[\frac{\partial^2}{\partial x \, \partial y} f(\mathbf{r} - \mathbf{a}^l) \right]_{\mathbf{r} = \mathbf{r}_{kk'}}$$

$$\times \exp[2\pi i(\mathbf{q} \cdot \mathbf{a}')] \exp[-2\pi i(\mathbf{q} \cdot \mathbf{r}_{kk'})]$$

$$= e_{k'}e_k \sum_l \left[\frac{\partial^2}{\partial x_{kk'} \, \partial y_{kk'}} f(\mathbf{r}_{kk'} - \mathbf{a}') \right]$$

$$\times \exp[2\pi i(\mathbf{q} \cdot \mathbf{a}')] \exp[-2\pi i(\mathbf{q} \cdot \mathbf{r}_{kk'})]$$

$$= e_k e_{k'} \exp[-2\pi i(\mathbf{q} \cdot \mathbf{r}_{kk'})] F^{\mathbf{q}}_{xy}(\mathbf{r}_{kk'}) \qquad (6.1)$$

and

$$^c\!\left[\begin{matrix} kk \\ xy \end{matrix} \right] = e_k^2 \lim_{\mathbf{r} \to 0} \left\{ \sum_l \frac{\partial^2}{\partial x \, \partial y} f(\mathbf{r} - \mathbf{a}') \exp[2\pi i(\mathbf{q} \cdot \mathbf{a}')] \right.$$

$$\left. - \frac{\partial^2}{\partial x \, \partial y} f(\mathbf{r}) \right\}$$

$$= e_k^2 \lim_{\mathbf{r} \to 0} [F^{\mathbf{q}}_{xy}(\mathbf{r}) - f_{xy}(\mathbf{r})] \qquad (6.2)$$

where

$$f_{xy}(\mathbf{r}) = \frac{\partial^2 f(\mathbf{r})}{\partial x \, \partial y}$$

We now employ Ewald's theta-function transformation, which shows that the function

$$\bar{F}^{\mathbf{q}}(\mathbf{r}) = 2\pi^{-1/2} \sum_l \exp[-\varepsilon^2 |\mathbf{r} - \mathbf{a}'|^2 + 2\pi i(\mathbf{q} \cdot \mathbf{a}')]$$

$$= \frac{2\pi}{v_a} \sum_h \varepsilon^{-3} \exp\left[-\frac{\pi^2}{\varepsilon^2} |\mathbf{b}_h + \mathbf{q}|^2 + 2\pi i(\mathbf{b}_h + \mathbf{q}) \cdot (\mathbf{r}) \right] \qquad (6.3)$$

where the first sum is over the direct lattice and the second sum is over the reciprocal lattice. It can readily be seen that

$$F^q(\mathbf{r}) = \int_0^\infty \bar{F}^q(\mathbf{r})\, d\varepsilon$$

$$= \int_0^E \left\{ \frac{2\pi}{v_a} \sum_h \varepsilon^{-3} \exp\left[-\frac{\pi^2}{\varepsilon^2}|\mathbf{b}_h + \mathbf{q}|^2 + 2\pi i(\mathbf{b}_h + \mathbf{q})\cdot(\mathbf{r}) \right] \right\} d\varepsilon$$

$$+ \int_E^\infty 2\pi^{-1/2} \sum_l \exp\{[-\varepsilon^2|\mathbf{r} - \mathbf{a}'|^2 + 2\pi i(\mathbf{q}\cdot\mathbf{a}')]\}\, d\varepsilon$$

which gives

$$F^q(\mathbf{r}) = (\pi v_a)^{-1} \sum_h |\mathbf{b}_h + \mathbf{q}|^{-2} \exp\left[-\frac{\pi^2}{E^2}|\mathbf{b}_h + \mathbf{q}|^2 + 2\pi i(\mathbf{b}_h + \mathbf{q})\cdot(\mathbf{r}) \right]$$

$$+ \sum_l \frac{1 - G(E|\mathbf{r} - \mathbf{a}'|)}{|\mathbf{r} - \mathbf{a}'|} \exp[2\pi i(\mathbf{q}\cdot\mathbf{a}')]$$

where

$$G(s) = 2\pi^{-1/2} \int_0^s \exp(-\xi^2)\, d\xi$$

Thus we obtain

$${}^c\!\begin{bmatrix} kk' \\ xy \end{bmatrix} = e_k e_{k'} \exp[-2\pi i(\mathbf{q}\cdot\mathbf{r}_{kk'})] F^q_{xy}(\mathbf{r}_{kk'}) \qquad (k \neq k')$$

where

$$F^q(\mathbf{r}_{kk'}) = (\pi v_a)^{-1} \sum_h |\mathbf{b}_h + \mathbf{q}|^{-2}$$

$$\times \exp\left[-\frac{\pi^2}{E^2}|\mathbf{b}_h + \mathbf{q}|^2 + 2\pi i(\mathbf{b}_h + \mathbf{q}) \cdot (\mathbf{r}_{kk'}) \right]$$

$$+ \sum_l \frac{1 - G(E|\mathbf{r}_{kk'} - \mathbf{a}^l|)}{|\mathbf{r}_{kk'} - \mathbf{a}^l|} \exp[2\pi i(\mathbf{q} \cdot \mathbf{a}^l)]$$

$$\frac{\partial F^q(\mathbf{r}_{kk'})}{\partial x_{kk'}} = \frac{2\pi i}{\pi v_a} \sum_h \frac{(b_{hx} + q_x)}{|\mathbf{b}_h + \mathbf{q}|^2} \exp\left[-\frac{\pi^2}{E^2}|\mathbf{b}_h + \mathbf{q}|^2 + 2\pi i(\mathbf{b}_h + \mathbf{q}) \cdot (\mathbf{r}_{kk'}) \right]$$

$$+ E \sum_l \frac{\partial \psi(E|\mathbf{r}_{kk'} - \mathbf{a}^l|)}{\partial x_{kk'}} \exp[2\pi i(\mathbf{q} \cdot \mathbf{a}^l)]$$

and

$$F^q_{xy}(\mathbf{r}_{kk'}) = \frac{\partial^2 F^q(\mathbf{r}_{kk'})}{\partial x_{kk'} \, \partial y_{kk'}}$$

$$= -\frac{4\pi}{v_a} \sum_h \frac{(b_{hx} + q_x)(b_{hy} + q_y)}{|\mathbf{b}_h + \mathbf{q}|^2}$$

$$\times \exp\left[-\frac{\pi^2}{E^2}|\mathbf{b}_h + \mathbf{q}|^2 + 2\pi i(\mathbf{b}_h + \mathbf{q}) \cdot (\mathbf{r}_{kk'}) \right]$$

$$+ E \sum_l \frac{\partial^2 \psi(E|\mathbf{r}^l_{kk'}|)}{\partial x_{kk'} \, \partial y_{kk'}} \exp[2\pi i(\mathbf{q} \cdot \mathbf{a}^l)]$$

Therefore

$$
{}^{c}\!\begin{bmatrix} kk' \\ xy \end{bmatrix} = e_k e_{k'} \left\{ -\frac{4\pi}{v_a} \sum_h \frac{(b_{hx} + q_x)(b_{hy} + q_y)}{|\mathbf{b}_h + \mathbf{q}|^2} \right.
$$

$$
\times \exp\!\left(-\frac{\pi^2}{E^2}|\mathbf{b}_h + \mathbf{q}|^2 - 2\pi i (\mathbf{b}_h \cdot \mathbf{r}_{k'k}) \right)
$$

$$
+ E^2 \sum_l \left[\psi'\!\left[E \Big| \mathbf{r}\!\begin{pmatrix} l \\ kk' \end{pmatrix} \Big| \right] \right] \frac{\delta_{xy}}{\Big| \mathbf{r}\!\begin{pmatrix} l \\ kk' \end{pmatrix} \Big|} + \left(E\psi''\!\left[E \Big| \mathbf{r}\!\begin{pmatrix} l \\ kk' \end{pmatrix} \Big| \right] \right.
$$

$$
\left. -\frac{\psi'\!\left[E \Big| \mathbf{r}\!\begin{pmatrix} l \\ kk' \end{pmatrix} \Big| \right]}{\Big| \mathbf{r}\!\begin{pmatrix} l \\ kk' \end{pmatrix} \Big|} \right) \frac{x\!\begin{pmatrix} l \\ kk' \end{pmatrix} y\!\begin{pmatrix} l \\ kk' \end{pmatrix}}{\Big| \mathbf{r}\!\begin{pmatrix} l \\ kk' \end{pmatrix} \Big|^2} \right] \exp\!\left\{ 2\pi i \left[\mathbf{q} \cdot \mathbf{r}\!\begin{pmatrix} l \\ k'k \end{pmatrix} \right] \right\}
$$

(6.4)

where

$$
\psi(s) = \frac{1 - G(s)}{s}
$$

$$
\psi'(s) = \frac{\partial \psi(s)}{\partial s}
$$

and

$$
\psi''(s) = \frac{\partial^2 \psi(s)}{\partial s^2}
$$

For the case $k = k'$, we have

$$
{}^{c}\!\begin{bmatrix} kk \\ xy \end{bmatrix} = e_k^2 \lim_{\mathbf{r} \to 0} \left[F_{xy}^{\mathbf{q}}(\mathbf{r}) - f_{xy}(\mathbf{r}) \right]
$$

Note that, as **r** approaches zero,

$$|\mathbf{r} - \mathbf{a}^l| \rightarrow |\mathbf{a}^l|$$

but

$$\mathbf{a}^l \equiv (0, 0, 0) \qquad \text{for } l = 0$$

Thus we must consider the $l = 0$ term separately. In this case,

$$\frac{\partial^2}{\partial x\, \partial y}\left[\frac{1 - G(E|\mathbf{r}|)}{|\mathbf{r}|}\right] - \frac{\partial^2}{\partial x\, \partial y}\frac{1}{|\mathbf{r}|}$$

$$= -\frac{\partial^2}{\partial x\, \partial y}\frac{G(E|\mathbf{r}|)}{|\mathbf{r}|}$$

$$= -\frac{\partial^2}{\partial x\, \partial y}\frac{1}{|\mathbf{r}|}\frac{2}{\pi^{1/2}}\int_0^{E|\mathbf{r}|} e^{-\xi^2}\, d\xi$$

$$= -\frac{2}{\pi^{1/2}}\frac{\partial^2}{\partial x\, \partial y}\frac{1}{|\mathbf{r}|}\int_0^{E|\mathbf{r}|}\left(1 - \xi^2 + \frac{\xi^4}{2!} - \frac{\xi^6}{3!} + \cdots\right) d\xi$$

$$= -\frac{2}{\pi^{1/2}}\frac{\partial^2}{\partial x\, \partial y}\frac{1}{|\mathbf{r}|}\left(E|\mathbf{r}| - \frac{E^3|\mathbf{r}|^3}{3} + \frac{E^5|\mathbf{r}|^5}{10} - \cdots\right)$$

$$= -\frac{2}{\pi^{1/2}}\frac{\partial^2}{\partial x\, \partial y}\left(E - \frac{E^3|\mathbf{r}|^2}{3} + \frac{E^5|\mathbf{r}|^4}{10} - \cdots\right)$$

$$= -\frac{2}{\pi^{1/2}}\left[-E^3\frac{2\delta_{xy}}{3} + E^5\frac{8xy(1 - \delta_{xy}) + 4|\mathbf{r}|^2\delta_{xy}}{10} - \cdots\right]$$

Therefore

$$\lim_{\mathbf{r}\to 0} = (4/3\pi^{1/2})E^3\delta_{xy}$$

We shall now employ this result in deriving

$$^c\begin{bmatrix} kk \\ xy \end{bmatrix} = e_k^2 \lim_{r \to 0} [F_{xy}^q(\mathbf{r}) - f_{xy}(\mathbf{r})]$$

As before,

$$F^q(\mathbf{r}) = (\pi v_a)^{-1} \sum_h |\mathbf{b}_h + \mathbf{q}|^{-2} \exp\left[-\frac{\pi^2}{E^2}|\mathbf{b}_h + \mathbf{q}|^2 + 2\pi i (\mathbf{b}_h + \mathbf{q}) \cdot (\mathbf{r}) \right]$$

$$+ \sum_l \frac{1 - G(E|\mathbf{r} - \mathbf{a}'|)}{|\mathbf{r} - \mathbf{a}'|} \exp[2\pi i (\mathbf{q} \cdot \mathbf{a}')]$$

$$\frac{\partial F^q(\mathbf{r})}{\partial x} = \frac{2\pi i}{\pi v_a} \sum_h \frac{(b_{hx} + q_x)}{|\mathbf{b}_h + \mathbf{q}|^2} \exp\left[-\frac{\pi^2}{E^2}|\mathbf{b}_h + \mathbf{q}|^2 + 2\pi i (\mathbf{b}_h + \mathbf{q}) \cdot (\mathbf{r}) \right]$$

$$+ \sum_l \frac{\partial}{\partial x} \frac{1 - G(E|\mathbf{r} - \mathbf{a}'|)}{|\mathbf{r} - \mathbf{a}'|} \exp[2\pi i (\mathbf{q} \cdot \mathbf{a}')]$$

and

$$\frac{\partial^2 F^q(\mathbf{r})}{\partial x\, \partial y} = -\frac{4\pi}{v_a} \sum_h \frac{(b_{hx} + q_x)(b_{hy} + q_y)}{|\mathbf{b}_h + \mathbf{q}|^2}$$

$$\times \exp\left[-\frac{\pi^2}{E^2}|\mathbf{b}_h + \mathbf{q}|^2 + 2\pi i (\mathbf{b}_h + \mathbf{q}) \cdot (\mathbf{r}) \right]$$

$$+ \sum_l \frac{\partial^2}{\partial x\, \partial y} \frac{1 - G(E|\mathbf{r} - \mathbf{a}'|)}{|\mathbf{r} - \mathbf{a}'|} \exp[2\pi i (\mathbf{q} \cdot \mathbf{a}')]$$

Therefore

$$
{}^c\!\left[\begin{matrix} kk \\ xy \end{matrix}\right] = e_k^2 \lim_{r\to 0} \frac{-4\pi}{v_a} \sum_h \frac{(b_{hx} + q_x)(b_{hy} + q_y)}{|\mathbf{b}_h + \mathbf{q}|^2}
$$

$$
\times \exp\!\left[-\frac{\pi^2}{E^2}|\mathbf{b}_h + \mathbf{q}|^2 + 2\pi i (\mathbf{b}_h + \mathbf{q}) \cdot (\mathbf{r}) \right]
$$

$$
+ e_k^2 \lim_{r\to 0} \left\{ \frac{\partial^2}{\partial x\,\partial y}\!\left[\frac{1 - G(E|\mathbf{r}|)}{|\mathbf{r}|} \right] - \frac{\partial^2}{\partial x\,\partial y} \frac{1}{|\mathbf{r}|} \right\}
$$

$$
+ e_k^2 \lim_{r\to 0} \left\{ E \sum_{l\neq 0} \frac{\partial^2 \psi(E|\mathbf{r} - \mathbf{a}'|)}{\partial x\,\partial y} \exp[2\pi i (\mathbf{q} \cdot \mathbf{a}')] \right\}
$$

Thus

$$
{}^c\!\left[\begin{matrix} kk \\ xy \end{matrix}\right] = e_k^2 \Bigg\{ \frac{-4\pi}{v_a} \sum_h \frac{(b_{hx} + q_x)(b_{hy} + q_y)}{|\mathbf{b}_h + \mathbf{q}|^2}
$$

$$
\times \exp\!\left(-\frac{\pi^2}{E^2}|\mathbf{b}_h + \mathbf{q}|^2 \right) + \frac{4}{3\pi^{1/2}} E^3 \delta_{xy}
$$

$$
+ E^2 \sum_{l\neq 0} \left[\psi'(E|\mathbf{a}'|)\frac{\delta_{xy}}{|\mathbf{a}'|} \right.
$$

$$
\left. + \left(E\psi''(E|\mathbf{a}'|) - \frac{\psi'(E|\mathbf{a}'|)}{|\mathbf{a}'|} \right)\frac{a_x' a_y'}{|\mathbf{a}'|^2} \right] \exp[2\pi i (\mathbf{q} \cdot \mathbf{a}')] \Bigg\} \quad (6.5)
$$

For reference purposes we show the Coulomb coupling coefficients calculated using Eqs. (6.4) and (6.5) for the sodium chloride and cesium chloride structures in Tables III and IV, respectively. These are tabulated for vectors within the irreducible one forty-eighth part of the first Brillouin zone, from which a uniform mesh of 1000 wave vectors within the full zone can be generated by symmetry. These values are accurate to 10^{-7} or better.

TABLE IIIa. Coulomb Coupling Coefficients $c\,[^{kk}_{\alpha\beta}]$ in Units of e^2/v_a for the NaCl-Structure Lattice: $1/10(P_x, P_y, P_z) = 2r_0(q_x, q_y, q_z)$

P_x	P_y	P_x	Weight	$c[^{11}_{xx}]$	$c[^{11}_{yy}]$	$c[^{11}_{zz}]$	$c[^{11}_{xy}]$	$c[^{11}_{xz}]$	$c[^{11}_{yz}]$
10	4	0	12	$-1.8133\ 4570$	$1.6320\ 6520$	$0.1812\ 8051$	$0.0000\ 0001$	$0.0000\ 0000$	$0.0000\ 0000$
10	2	2	12	$-2.9381\ 9090$	$1.4690\ 9540$	$1.4690\ 9540$	$-0.0000\ 0001$	$-0.0000\ 0001$	$-1.0415\ 8370$
10	2	2	12	$-3.6033\ 7250$	$1.9617\ 7910$	$1.6415\ 9340$	$-0.0000\ 0001$	$0.0000\ 0000$	$0.0000\ 0000$
10	0	0	3	$-4.3338\ 6560$	$2.1669\ 3280$	$2.1669\ 3280$	$0.0000\ 0000$	$0.0000\ 0000$	$0.0000\ 0000$
9	5	1	24	$-0.7176\ 7910$	$1.4353\ 5820$	$-0.7176\ 7907$	$-1.0072\ 1580$	$-0.3206\ 9634$	$-1.0072\ 1580$
9	3	3	12	$-1.5849\ 7300$	$0.7924\ 8650$	$0.7924\ 8650$	$-0.8520\ 1392$	$-0.8520\ 1392$	$-2.1041\ 0610$
9	3	1	48	$-2.6972\ 2940$	$1.6595\ 5940$	$1.0376\ 6990$	$-0.8800\ 4706$	$-0.3335\ 9867$	$-0.7642\ 6415$
9	1	1	24	$-4.0518\ 9900$	$2.0259\ 4950$	$2.0259\ 4950$	$-0.3548\ 0687$	$-0.3548\ 0687$	$-0.2824\ 1390$
8	6	0	24	$0.0577\ 2618$	$1.2439\ 2840$	$-1.3016\ 5470$	$-1.7997\ 7080$	$0.0000\ 0000$	$0.0000\ 0000$
8	4	2	48	$-1.4726\ 7450$	$1.1352\ 5890$	$0.3374\ 1564$	$-1.9420\ 3880$	$-1.1981\ 2800$	$-1.8175\ 3440$
8	4	0	24	$-1.9758\ 0330$	$1.4427\ 7300$	$0.5330\ 3029$	$-1.9811\ 1130$	$0.0000\ 0000$	$0.0000\ 0000$
8	2	2	24	$-3.1892\ 6910$	$1.5946\ 3450$	$1.5946\ 3450$	$-1.2853\ 7520$	$-1.2853\ 7520$	$-1.0808\ 1160$
8	2	0	24	$-3.9160\ 4040$	$2.0322\ 6920$	$1.8837\ 7130$	$-1.3405\ 0490$	$0.0000\ 0000$	$0.0000\ 0000$
8	0	0	6	$-4.7425\ 3480$	$2.3712\ 6730$	$2.3712\ 6730$	$0.0000\ 0000$	$0.0000\ 0000$	$0.0000\ 0000$
7	7	1	12	$0.7924\ 8644$	$0.7924\ 8644$	$-1.5849\ 7290$	$-2.1041\ 0620$	$-0.8520\ 1391$	$-0.8520\ 1391$
7	5	3	24	$-0.2857\ 4696$	$0.5714\ 9380$	$-0.2857\ 4690$	$-2.8094\ 3620$	$-2.2988\ 5430$	$-2.8094\ 3620$
7	5	1	48	$-0.8948\ 9530$	$0.8386\ 0513$	$0.0562\ 9014$	$-2.8675\ 9900$	$-0.8685\ 0975$	$-1.0283\ 5360$
7	3	3	24	$-1.8577\ 2470$	$0.9288\ 6235$	$0.9288\ 6235$	$-2.4178\ 9400$	$-2.4178\ 9400$	$-2.1940\ 4700$
7	3	1	48	$-3.0503\ 7710$	$1.5264\ 2820$	$1.5239\ 4880$	$-2.6165\ 9290$	$-0.9725\ 1017$	$-0.8312\ 5331$
7	1	1	24	$-4.7053\ 5820$	$2.3526\ 7910$	$2.3526\ 7910$	$-1.0909\ 2980$	$-1.0909\ 2980$	$-0.3216\ 6739$
6	6	2	24	$0.1385\ 7654$	$0.1385\ 7654$	$-0.2771\ 5313$	$-3.1765\ 2780$	$-1.9343\ 3930$	$-1.9343\ 3930$
6	6	0	12	$0.0148\ 7011$	$0.0148\ 7011$	$-0.0297\ 4023$	$-3.2345\ 1140$	$0.0000\ 0000$	$0.0000\ 0000$
6	4	4	24	$-0.5532\ 8645$	$0.2766\ 4319$	$0.2766\ 4319$	$-3.2948\ 8150$	$-3.2948\ 8150$	$-3.2032\ 9060$
6	4	2	48	$-1.5942\ 2670$	$0.6220\ 6900$	$0.9721\ 5764$	$-3.6033\ 6500$	$-2.1420\ 0560$	$-1.9899\ 5740$

TABLE IIIa (*contd.*)

P_x	P_y	P_z	Weight	$c^{[11]}_{xx}$	$c^{[11]}_{yy}$	$c^{[11]}_{zz}$	$c^{[11]}_{xy}$	$c^{[11]}_{xz}$	$c^{[11]}_{yz}$
6	4	0	24	−2.0963 7780	0.6776 7274	1.4187 0510	−3.8017 2170	0.0000 0000	0.0000 0000
6	2	2	24	−3.5977 2260	1.7988 6130	1.7988 6130	−2.5422 2240	−2.5422 2240	−1.2980 3430
6	2	0	24	−4.5684 1920	2.0676 6100	2.5007 5810	−2.7696 6500	0.0000 0000	0.0000 0000
6	0	0	6	−5.7892 4450	2.8946 2220	2.8946 2220	0.0000 0000	0.0000 0000	0.0000 0000
5	5	5	4	−0.0000 0001	−0.0000 0001	−0.0000 0001	−3.6151 4720	−3.6151 4720	−3.6151 4720
5	5	3	24	−0.2242 6553	−0.2242 6553	0.4485 3100	−3.8327 1100	−2.9799 2530	−2.9799 2530
5	5	1	24	−0.6025 7404	−0.6025 7404	1.2051 4800	−4.2001 7400	−1.1738 3240	−1.1738 3240
5	3	3	24	−1.6605 1490	0.8302 5744	0.8302 5744	−3.5790 0770	−3.5790 0770	−2.6453 0980
5	3	1	48	−3.0535 2980	0.8337 2618	2.2198 0350	−4.2695 5100	−1.5378 9990	−1.1021 4840
5	1	1	24	−5.5320 6920	2.7660 3450	2.7660 3450	−1.9625 6510	−1.9625 6510	−0.4827 0393
4	4	4	8	−0.0000 0001	−0.0000 0001	−0.0000 0001	−3.6693 6880	−3.6693 6880	−3.6693 6880
4	4	2	24	−0.7498 0442	−0.7498 0442	1.4996 0880	−4.5610 1290	−2.5912 9840	−2.5912 9840
4	4	0	12	−1.2237 8070	−1.2237 8070	2.4475 6120	−5.0865 0060	0.0000 0000	0.0000 0000
4	2	2	24	−3.1189 0620	1.5594 5300	1.5594 5300	−3.7200 1420	−3.7200 1420	−2.0390 5290
4	2	0	24	−4.6462 7300	1.4155 2720	3.2307 4560	−4.4648 3960	0.0000 0000	0.0000 0000
4	0	0	6	−7.0380 1310	3.5190 0650	3.5190 0650	0.0000 0000	0.0000 0000	0.0000 0000
3	3	3	8	−0.0000 0003	−0.0000 0003	−0.0000 0003	−3.8118 8680	−3.8118 8680	−3.8118 8680
3	3	1	24	−1.3530 1660	−1.3530 1660	2.7060 3300	−5.3647 7610	−1.8743 2250	−1.8743 2250
3	1	1	24	−5.3782 5410	2.6891 2700	2.6891 2700	−3.2440 2420	−3.2440 2420	−1.1127 5370
2	2	2	8	−0.0000 0006	−0.0000 0006	−0.0000 0006	−3.9891 6180	−3.9891 6180	−3.9891 6180
2	2	0	12	−1.8981 2590	−1.8981 2590	3.7962 5170	−6.0330 1560	0.0000 0000	0.0000 0000
2	0	0	6	−8.0131 6070	4.0065 8020	4.0065 8020	0.0000 0000	0.0000 0000	0.0000 0000
1	1	1	8	−0.0000 0007	−0.0000 0007	−0.0000 0007	−4.1334 6520	−4.1334 6520	−4.1334 6520

TABLE IIIb. Coulomb Coupling Coefficients $c^{[kk']}_{\alpha\beta}$ in Units of e^2/v_a for the NaCl-Structure Lattice:

$$1/10(P_x, P_y, P_z) = 2r_0(q_x, q_y, q_z)$$

P_x	P_y	P_z	Weight	$c^{[12]}_{xx}$	$c^{[12]}_{yy}$	$c^{[12]}_{zz}$	$c^{[12]}_{xy}$	$c^{[12]}_{xz}$	$c^{[12]}_{yz}$
10	4	0	12	12.1690 0600	-2.3097 3360	-9.8592 7340	0.0000 0000	0.0000 0000	0.0000 0000
10	2	2	12	13.4486 5300	-6.7243 2650	-6.7243 2650	0.0000 0000	0.0000 0000	-0.3935 5338
10	2	0	12	14.2134 6100	-6.0699 2920	-8.1435 3270	0.0000 0000	0.0000 0000	0.0000 0000
10	0	0	3	15.0410 2100	-7.5205 1020	-7.5205 1020	0.0000 0000	0.0000 0000	0.0000 0000
9	5	1	24	10.5334 3800	0.0000 0000	-10.5334 3800	0.4362 5715	0.0000 0000	-0.4362 5713
9	3	3	12	11.5141 5000	-5.7570 7550	-5.7570 7550	0.2517 0589	0.2517 0589	-0.8580 2849
9	3	1	48	12.8610 1300	-4.4366 6000	-8.4243 5380	0.4484 7926	0.1197 1623	-0.2680 1965
9	1	1	24	14.4282 8800	-7.2141 4420	-7.2141 4420	0.1964 8134	0.1964 8134	-0.0838 5847
8	6	0	24	8.5502 2140	3.0007 1980	-11.5509 4100	0.8048 5725	0.0000 0000	0.0000 0000
8	4	2	48	10.3754 0200	-2.3723 5470	-8.0030 4780	0.8081 3427	0.2372 6819	-0.5721 4102
8	4	0	24	11.0801 7700	-1.6874 4940	-9.3927 2830	1.0465 5680	0.0000 0000	0.0000 0000
8	2	2	24	12.5236 1700	-6.2618 0860	-6.2618 0860	0.6324 2877	0.6324 2877	-0.2453 4676
8	2	0	24	13.4106 2300	-5.6309 4240	-7.7796 8150	0.7905 0223	0.0000 0000	0.0000 0000
8	0	0	6	14.3837 1800	-7.1918 5950	-7.1918 5950	0.0000 0000	0.0000 0000	0.0000 0000
7	7	1	12	5.7570 7540	5.7570 7540	-11.5141 5000	0.8580 2854	-0.2517 0588	-0.2517 0588
7	5	3	24	6.8346 3880	0.0000 0006	-6.8346 3890	0.7431 6084	0.0000 0001	-0.7431 6077
7	5	1	48	8.1457 8300	1.3362 6190	-9.4820 4500	1.4242 4540	0.0747 7454	-0.1895 0246
7	3	3	24	9.2784 3000	-4.6392 1490	-4.6392 1490	0.9172 0340	0.9172 0340	-0.3113 6835
7	3	1	48	10.9679 7400	-3.3631 6380	-7.6048 1070	1.5192 3800	0.4338 1830	-0.0520 9590
7	1	1	24	13.0198 5500	-6.5099 2800	-6.5099 2800	0.6870 5326	0.6870 5326	0.0056 5677
6	6	2	24	4.3075 3410	4.3075 3410	-8.6150 6830	1.2948 6520	-0.1722 3571	-0.1722 3571
6	6	0	12	5.0000 3930	5.0000 3930	-10.0000 7800	1.6975 6370	0.0000 0000	0.0000 0000
6	4	4	24	4.9298 3040	-2.4649 1520	-2.4649 1520	0.6929 8579	0.6929 8579	-0.2078 4157
6	4	2	48	7.1154 1620	-0.5492 5507	-5.5661 6110	1.8661 9630	0.6920 0436	0.1122 4489

TABLE IIIb. (*contd.*)

P_x	P_y	P_z	Weight	$c^{[12]}_{[xx]}$	$c^{[12]}_{[yy]}$	$c^{[12]}_{[zz]}$	$c^{[12]}_{[xy]}$	$c^{[12]}_{[xz]}$	$c^{[12]}_{[yz]}$
6	4	0	24	8.0219 1840	0.1840 0124	-8.2059 1960	2.3480 0320	0.0000 0000	0.0000 0000
6	2	2	24	9.9312 6410	-4.9656 3200	-4.9656 3200	1.5324 0060	1.5324 0060	0.2364 1324
6	2	0	24	11.2004 5200	-4.3570 1660	-6.8434 3530	1.9097 7480	0.0000 0000	0.0000 0000
6	0	0	6	12.6845 6000	-6.3422 8000	-6.3422 8000	0.0000 0000	0.0000 0000	0.0000 0000
5	5	5	4	-0.0000 0001	-0.0000 0001	-0.0000 0001	0.0000 0000	0.0000 0000	0.0000 0000
5	5	3	24	2.3782 0540	2.3782 0540	-4.7564 1090	1.5056 3610	0.4345 7332	0.4345 7332
5	5	1	24	3.9238 2480	3.9238 2480	-7.8476 4960	2.5118 3460	0.2758 4361	0.2758 4361
5	3	3	24	5.2479 4880	-2.6239 7430	-2.6239 7430	1.8534 6520	1.8534 6520	0.7891 7267
5	3	1	48	7.4653 6840	-1.1972 1460	-6.2681 5360	2.9747 6140	0.9117 4088	0.4354 1914
5	1	1	24	10.5927 3000	-5.2963 6530	-5.2963 6530	1.4837 0020	1.4837 0020	0.2408 9350
4	4	4	8	0.0000 0001	0.0000 0001	0.0000 0001	1.3290 7720	1.3290 7720	1.3290 7720
4	4	2	24	2.3504 5190	2.3504 5190	-4.7009 0370	2.9333 1460	1.2571 6950	1.2571 6950
4	4	0	12	3.4018 2270	3.4018 2270	-6.8036 4530	3.7006 7850	0.0000 0000	0.0000 0000
4	2	2	24	5.8903 4180	-2.9451 7080	-2.9451 7080	2.7639 4670	2.7639 4670	1.2665 9480
4	2	0	24	7.8509 2020	-2.1335 3380	-5.7173 8620	3.6403 1730	0.0000 0000	0.0000 0000
4	0	0	6	10.6262 6100	-5.3131 3050	-5.3131 3050	0.0000 0000	0.0000 0000	0.0000 0000
3	3	3	8	0.0000 0005	0.0000 0005	0.0000 0005	2.5097 6470	2.5097 6470	2.5097 6470
3	3	1	24	2.4071 0910	2.4071 0910	-4.8142 1810	4.3639 9250	1.3973 1150	1.3973 1150
3	1	1	24	7.2824 5740	-3.6412 2860	-3.6412 2860	2.8722 3820	2.8722 3820	0.9370 8701
2	2	2	8	0.0000 0006	0.0000 0006	0.0000 0006	3.4233 6900	3.4233 6900	3.4233 6900
2	2	0	12	2.4151 4590	2.4151 4590	-4.8302 9160	5.5402 9210	0.0000 0000	0.0000 0000
2	0	0	6	8.9937 0040	-4.4968 5010	-4.4968 5010	0.0000 0000	0.0000 0000	0.0000 0000
1	1	1	8	0.0000 0010	0.0000 0010	0.0000 0010	3.9949 9240	3.9949 9240	3.9949 9240

TABLE IVa. Coulomb Coupling Coefficients $c_{[\alpha\beta]}^{[kk]}$ in Units of e^2/v_a for the CsCl-Structure Lattice: $\frac{1}{10}(P_x, P_y, P_z) = r_0(q_x, q_y, q_z)$ [a]

P_x	P_y	P_z	Weight	$c_{[xx]}^{[11]}$	$c_{[yy]}^{[11]}$	$c_{[zz]}^{[11]}$	$c_{[xy]}^{[11]}$	$c_{[xz]}^{[11]}$	$c_{[yz]}^{[11]}$
5	5	5	1	0.0000 0000	0.0000 0000	0.0000 0000	0.0000 0000	0.0000 0000	0.0000 0000
5	5	4	6	−0.2451 3495	−0.2451 3495	0.4902 6990	0.0000 0000	0.0000 0000	0.0000 0000
5	5	3	6	−0.8970 6206	−0.8970 6206	1.7941 2390	0.0000 0000	0.0000 0000	0.0000 0000
5	5	2	6	−1.7238 2890	−1.7238 2890	3.4476 5780	0.0000 0000	0.0000 0000	0.0000 0000
5	5	1	6	−2.4102 9610	−2.4102 9610	4.8205 9210	0.0000 0000	0.0000 0000	0.0000 0000
5	5	0	3	−2.6767 8860	−2.6767 8860	5.3535 7730	0.0000 0000	0.0000 0000	0.0000 0000
5	4	4	12	−0.5170 1999	0.2585 0999	0.2585 0999	0.0000 0000	0.0000 0000	−0.2463 6175
5	4	3	24	−1.2433 2030	−0.3590 0331	1.6023 2360	0.0000 0000	0.0000 0000	−0.4122 6117
5	4	2	24	−2.1713 3860	−1.1446 7780	3.3160 1640	0.0000 0000	0.0000 0000	−0.4299 4515
5	4	1	24	−2.9479 5510	−1.7993 3660	4.7472 9170	0.0000 0000	0.0000 0000	−0.2750 0137
5	4	0	12	−3.2509 6960	−2.0540 7500	5.3050 4470	0.0000 0000	0.0000 0000	0.0000 0000
5	3	3	12	−2.1780 4200	1.0890 2100	1.0890 2100	0.0000 0000	0.0000 0000	−0.6929 1108
5	3	2	24	−3.3936 0740	0.4308 3704	2.9627 7030	0.0000 0000	0.0000 0000	−0.7268 5923
5	3	1	24	−4.4298 4910	−0.1223 3818	4.5521 8720	0.0000 0000	0.0000 0000	−0.4672 7725
5	3	0	12	−4.8389 9650	−0.3388 3411	5.1778 3060	0.0000 0000	0.0000 0000	0.0000 0000
5	2	2	12	−5.0149 0950	2.5074 5480	2.5074 5480	0.0000 0000	0.0000 0000	−0.7684 7389
5	2	1	24	−6.4262 9810	2.1223 2430	4.3039 7380	0.0000 0000	0.0000 0000	−0.4974 5678
5	2	0	12	−6.9911 7630	1.9708 9940	5.0202 7690	0.0000 0000	0.0000 0000	0.0000 0000
5	1	1	12	−8.1934 2200	4.0967 1100	4.0967 1100	0.0000 0000	0.0000 0000	−0.3240 1519
5	1	0	12	−8.9084 1720	4.0158 5410	4.8925 6300	0.0000 0000	0.0000 0000	0.0000 0000
5	0	0	3	−9.6874 4300	4.8437 2140	4.8437 2140	0.0000 0000	0.0000 0000	0.0000 0000
4	4	4	8	0.0000 0000	0.0000 0000	0.0000 0000	−0.2828 9638	−0.2828 9638	−0.2828 9638
4	4	3	24	−0.6928 5893	−0.6928 5893	1.3857 1780	−0.3862 2906	−0.4780 3377	−0.4780 3377
4	4	2	24	−1.5833 8540	−1.5833 8540	3.1667 7080	−0.5308 1060	−0.5049 1086	−0.5049 1086

[a] r_0 denotes the cube cell side (i.e., the Cs^+–Cs^+ separation) for these structures.

TABLE IVa (*contd.*)

P_x	P_y	P_z	Weight	$c^{[11]}_{xx}$	$c^{[11]}_{yy}$	$c^{[11]}_{zz}$	$c^{[11]}_{xy}$	$c^{[11]}_{zx}$	$c^{[11]}_{yz}$
4	4	1	24	-2.3335 8700	-2.3335 8700	4.6671 7410	-0.6630 7925	-0.3264 7324	-0.3264 7324
4	4	0	12	-2.6276 1560	-2.6276 1560	5.2552 3100	-0.7175 6861	0.0000 0000	0.0000 0000
4	3	3	24	-1.6006 4760	0.8003 2378	0.8003 2378	-0.6670 6449	-0.6670 6449	-0.8138 4918
4	3	2	48	-2.7970 0180	0.0364 0699	2.7605 9480	-0.9388 5629	-0.7250 0065	-0.8685 8438
4	3	1	48	-3.8328 1610	-0.6185 4786	4.4513 6390	-1.1948 3780	-0.4804 2992	-0.5669 5227
4	3	0	24	-4.2462 4760	-0.8783 9574	5.1246 4320	-1.3023 1400	0.0000 0000	0.0000 0000
4	2	2	24	-4.4461 1080	2.2230 5540	2.2230 5540	-1.0532 8750	-1.0532 8750	-0.9408 3135
4	2	1	48	-5.9240 3850	1.7538 0680	4.1702 3170	-1.3750 8650	-0.7176 9810	-0.6226 9673
4	2	0	24	-6.5279 9050	1.5651 1120	4.9628 7920	-1.5138 3400	0.0000 0000	0.0000 0000
4	1	1	24	-7.8564 4320	3.9282 2160	3.9282 2160	-0.9589 0199	-0.9589 0199	-0.4177 3245
4	1	0	24	-8.6633 3220	3.8316 0580	4.8317 2630	-1.0655 0350	0.0000 0000	0.0000 0000
4	0	0	6	-9.5631 2690	4.7815 6340	4.7815 6340	0.0000 0000	0.0000 0000	0.0000 0000
3	3	3	8	0.0000 0000	0.0000 0000	0.0000 0000	-1.1701 4580	-1.1701 4580	-1.1701 4580
3	3	2	24	-1.0941 3600	-1.0941 3600	2.1882 7200	-1.7055 6610	-1.3009 4790	-1.3009 4790
3	3	1	24	-2.0844 7870	-2.0844 7870	4.1689 5750	-2.2356 9650	-0.8810 5181	-0.8810 5181
3	3	0	12	-2.4925 8460	-2.4925 8460	4.9851 6920	-2.4660 3750	0.0000 0000	0.0000 0000
3	2	2	24	-2.7415 5840	1.3707 7920	1.3707 7920	-1.9939 3360	-1.9939 3360	-1.4977 2450
3	2	1	48	-4.3548 2140	0.5856 6201	3.7691 5930	-2.7347 8070	-1.4170 0490	-1.0511 0110

TABLE IVa (*contd.*)

P_x	P_y	P_z	Weight	$c^{[11]}_{xx}$	$c^{[11]}_{yy}$	$c^{[11]}_{zz}$	$c^{[11]}_{xy}$	$c^{[11]}_{zx}$	$c^{[11]}_{yz}$
3	2	0	24	−5.0591 4810	0.2468 3577	4.8123 1230	−3.0748 6240	0.0000 0000	0.0000 0000
3	1	1	24	−6.7644 9330	3.3822 4660	3.3822 4660	−2.0373 1150	−2.0373 1150	−0.7672 2381
3	1	0	24	−7.8844 3570	3.2123 3880	4.6720 9670	−2.3397 1990	0.0000 0000	0.0000 0000
3	0	0	6	−9.2369 0220	4.6184 5100	4.6184 5100	0.0000 0000	0.0000 0000	0.0000 0000
2	2	2	8	0.0000 0000	0.0000 0000	0.0000 0000	−2.4992 2290	−2.4992 2290	−2.4992 2290
2	2	1	24	−1.5501 2820	−1.5501 2820	3.1002 5620	−3.7471 6370	−1.9242 3390	−1.9242 3390
2	2	0	12	−2.3128 0160	−2.3128 0160	4.6256 0330	−4.3935 8950	0.0000 0000	0.0000 0000
2	1	1	24	−4.5046 2400	2.2523 1190	2.2523 1190	−3.2419 8040	−3.2419 8040	−1.6528 2380
2	1	0	24	−6.2485 9650	1.7745 3050	4.4740 6590	−4.0525 7840	0.0000 0000	0.0000 0000
2	0	0	6	−8.8321 3700	4.4160 6840	4.4160 6840	0.0000 0000	0.0000 0000	0.0000 0000
1	1	1	8	0.0000 0000	0.0000 0000	0.0000 0000	−3.7062 6530	−3.7062 6530	−3.7062 6530
1	1	0	12	−2.1566 3580	−2.1566 3580	4.3132 7160	−5.7866 5380	0.0000 0000	0.0000 0000
1	0	0	6	−8.5034 3030	4.2517 1510	4.2517 1510	0.0000 0000	0.0000 0000	0.0000 0000

TABLE IVb. Coulomb Coupling Coefficients $c^{[kk']}_{\alpha\beta}$ in Units of e^2/v_a for the CsCl-Structure Lattice: $1/10(P_x, P_y, P_z) = r_0(q_x, q_y, q_z)$

P_x	P_y	P_z	Weight	$c^{[12]}_{xx}$	$c^{[12]}_{yy}$	$c^{[12]}_{zz}$	$c^{[12]}_{xy}$	$c^{[12]}_{xz}$	$c^{[12]}_{yz}$
5	5	5	1	0.0000 0000	0.0000 0000	0.0000 0000	0.0000 0000	0.0000 0000	0.0000 0000
5	5	4	6	0.0000 0000	0.0000 0000	0.0000 0000	3.1239 0390	0.0000 0000	0.0000 0000
5	5	3	6	0.0000 0000	0.0000 0000	0.0000 0000	6.0225 4440	0.0000 0000	0.0000 0000
5	5	2	6	0.0000 0000	0.0000 0000	0.0000 0000	8.4299 5170	0.0000 0000	0.0000 0000
5	5	1	6	0.0000 0000	0.0000 0000	0.0000 0000	10.0473 3800	0.0000 0000	0.0000 0000
5	5	0	3	0.0000 0000	0.0000 0000	0.0000 0000	10.6204 7200	0.0000 0000	0.0000 0000
5	4	4	12	0.0000 0000	0.0000 0000	0.0000 0000	3.0186 3120	3.0186 3120	0.0000 0000
5	4	3	24	0.0000 0000	0.0000 0000	0.0000 0000	5.8293 1040	2.6770 2080	0.0000 0000
5	4	2	24	0.0000 0000	0.0000 0000	0.0000 0000	8.1774 0810	2.0478 8680	0.0000 0000
5	4	1	24	0.0000 0000	0.0000 0000	0.0000 0000	9.7647 8060	1.1225 9800	0.0000 0000
5	4	0	12	0.0000 0000	0.0000 0000	0.0000 0000	10.3295 4400	0.0000 0000	0.0000 0000
5	3	3	12	0.0000 0000	0.0000 0000	0.0000 0000	5.1932 5200	5.1932 5200	0.0000 0000
5	3	2	24	0.0000 0000	0.0000 0000	0.0000 0000	7.3292 8570	3.9968 5120	0.0000 0000
5	3	1	24	0.0000 0000	0.0000 0000	0.0000 0000	8.7980 1330	2.2024 8660	0.0000 0000
5	3	0	12	0.0000 0000	0.0000 0000	0.0000 0000	9.3263 4100	0.0000 0000	0.0000 0000
5	2	2	12	0.0000 0000	0.0000 0000	0.0000 0000	5.6866 8820	5.6866 8820	0.0000 0000
5	2	1	24	0.0000 0000	0.0000 0000	0.0000 0000	6.8750 4200	3.1560 8120	0.0000 0000
5	2	0	12	0.0000 0000	0.0000 0000	0.0000 0000	7.3088 0590	0.0000 0000	0.0000 0000
5	1	1	12	0.0000 0000	0.0000 0000	0.0000 0000	3.8399 1460	3.8399 1460	0.0000 0000
5	1	0	12	0.0000 0000	0.0000 0000	0.0000 0000	4.0927 2370	0.0000 0000	0.0000 0000
5	0	0	3	0.0000 0000	0.0000 0000	0.0000 0000	0.0000 0000	0.0000 0000	0.0000 0000
4	4	4	8	0.0000 0000	0.0000 0000	0.0000 0000	2.9228 9030	2.9228 9030	2.9228 9030
4	4	3	24	0.0282 0114	0.0282 0114	-0.0564 0228	5.6572 8980	2.6067 4160	2.6067 4160
4	4	2	24	0.0918 2497	0.0918 2497	-0.1836 4995	7.9605 7220	2.0091 6470	2.0091 6470

TABLE IVb (contd.)

P_x	P_y	P_z	Weight	$c^{[12]}_{xx}$	$c^{[12]}_{yy}$	$c^{[12]}_{zz}$	$c^{[12]}_{xy}$	$c^{[12]}_{xz}$	$c^{[12]}_{yz}$
4	4	1	24	0.1636 8805	0.1636 8805	−0.3273 7611	9.5318 2790	1.1086 3190	1.1086 3190
4	4	0	12	0.1959 4228	0.1959 4228	−0.3918 8457	10.0942 4800	0.0000 0000	0.0000 0000
4	3	3	24	0.1252 4872	−0.0626 2437	−0.0626 2437	5.0777 9250	5.0777 9250	2.3611 3430
4	3	2	48	0.3112 4376	−0.0001 2682	−0.3111 1696	7.2083 2530	3.9487 3940	1.8582 4400
4	3	1	48	0.5168 1317	0.0928 3512	−0.6096 4832	8.6999 4570	2.1966 2000	1.0444 0730
4	3	0	24	0.6093 3782	0.1380 8065	−0.7474 1850	9.2432 8510	0.0000 0000	0.0000 0000
4	2	2	24	0.7037 8701	−0.3518 9352	−0.3518 9352	5.6766 2250	5.6766 2250	1.5047 1800
4	2	1	48	1.1396 6500	−0.3279 1509	−0.8117 5000	6.9322 5990	3.1955 1560	0.8676 0517
4	2	0	24	1.3386 3740	−0.3064 7361	−1.0321 6380	7.4015 0590	0.0000 0000	0.0000 0000
4	1	1	24	1.8439 4570	−0.9219 7289	−0.9219 7289	3.9474 4870	3.9474 4870	0.5121 1491
4	1	0	24	2.1711 7850	−0.9545 2376	−1.2166 5490	4.2355 0760	0.0000 0000	0.0000 0000
4	0	0	6	2.5611 6800	−1.2805 8400	−1.2805 8400	0.0000 0000	0.0000 0000	0.0000 0000
3	3	3	8	−0.0000 0001	−0.0000 0001	−0.0000 0001	4.6555 3980	4.6555 3980	4.6555 3980
3	3	2	24	0.2413 0633	0.2413 0633	−0.4826 1271	6.7275 8270	3.7312 2550	3.7312 2550
3	3	1	24	0.5593 7378	0.5593 7378	−1.1187 4760	8.2604 9690	2.1349 3160	2.1349 3160
3	3	0	12	0.7128 0682	0.7128 0682	−1.4256 1370	8.8411 4790	0.0000 0000	0.0000 0000
3	2	2	24	0.8560 7379	−0.4280 3692	−0.4280 3692	5.5467 4280	5.5467 4280	3.1269 7250
3	2	1	48	1.6554 0920	−0.2112 6044	−1.4441 4880	7.0133 6440	3.2698 2220	1.8695 8840

TABLE IVb (*contd.*)

P_x	P_y	P_z	Weight	$c_{[xx]}^{[12]}$	$c_{[yy]}^{[12]}$	$c_{[zz]}^{[12]}$	$c_{[xy]}^{[12]}$	$c_{[xz]}^{[12]}$	$c_{[yz]}^{[12]}$
3	2	0	24	2.0514 3510	−0.0825 2048	−1.9689 1470	7.6059 6470	0.0000 0000	0.0000 0000
3	1	1	24	3.1359 2140	−1.5679 6080	−1.5679 6080	4.2757 9030	4.2757 9030	1.1708 6260
3	1	0	24	3.9032 9950	−1.5822 8180	−2.3210 1770	4.7122 1100	0.0000 0000	0.0000 0000
3	0	0	6	4.8860 9280	−2.4430 4650	−2.4430 4650	0.0000 0000	0.0000 0000	0.0000 0000
2	2	2	8	0.0000 0000	0.0000 0000	0.0000 0000	4.9335 7320	4.9335 7320	4.9335 7320
2	2	1	24	0.8535 4437	0.8535 4437	−1.7070 8880	6.6923 3140	3.1679 5500	3.1679 5500
2	2	0	12	1.3597 9070	1.3597 9070	−2.7195 8160	7.5148 4980	0.0000 0000	0.0000 0000
2	1	1	24	3.0171 1130	−1.5085 5560	−1.5085 5560	4.7527 1760	4.7527 1760	2.2800 0340
2	1	0	24	4.4700 7000	−1.2638 4130	−3.2062 2870	5.6549 3870	0.0000 0000	0.0000 0000
2	0	0	6	6.7498 3940	−3.3749 1970	−3.3749 1970	0.0000 0000	0.0000 0000	0.0000 0000
1	1	1	8	−0.0000 0004	−0.0000 0004	−0.0000 0004	4.4718 1340	4.4718 1340	4.4718 1340
1	1	0	12	1.8901 2360	1.8901 2360	−3.7802 4740	6.5993 6610	0.0000 0000	0.0000 0000
1	0	0	6	7.9585 1350	−3.9792 5670	−3.9792 5670	0.0000 0000	0.0000 0000	0.0000 0000

In addition to its computational utility, the advantage of the theta-function transformation is that it allows explicit separation of the first term in the electric field equation. This is the field that would be associated with the dipolar wave when one uses the first Maxwell equation and its coefficient is given by the $h = 0$ term in Eqs. (6.4) and (6.5).

This term is apparently not well defined in the limit of $\mathbf{q} \to 0$. The reason for this is very simple. In this limit the effect of surface polarization of the crystal is crucial, and the values of the dipolar coupling coefficients then depend on the geometry of the specimen. Finally, in the derivation just given, we have neglected the retardation of the electromagnetic field. This modifies the situation in an infinite crystal in the way described in Section 4.

In order to see clearly where the theta-function transformation breaks down for a finite crystal, let us examine the equations given above. In making the theta-function transformation [i.e., Eq. (6.3)] it is implicit that the sum over l can be approximated by the sum over an infinite crystal. As the wavelength ($|\mathbf{q}|^{-1}$) becomes comparable to the size of the crystal, this ceases to be true, and the theta-function transformation breaks down. In these circumstances the elements of $\mathbf{H}$ cease to be well defined, and one must include the effect of surface polarization. The fraction of modes affected by this is on the order of the number of surface sites divided by the number of internal sites and is therefore negligible in the calculation of integrated properties. The total error is comparable to that introduced by the use of periodic boundary conditions for the remaining modes; the difference is that between a large effect for a few modes and a small effect for all the modes.

In a finite crystal there come into existence a whole variety of rather complex electromagnetic surface modes. The theoretical discussion of such modes is a more difficult problem because of the breakdown of the translational invariance normal to the surface. Thus three-dimensional plane-wave solutions of the equations of motion are not appropriate. A full discussion of calculations on surface modes in ionic and nonionic crystals is given in a review by Maradudin *et al.*[104]

In discussing surface modes it is often convenient to utilize somewhat different techniques for transforming the dipolar sums.[105] These are more restricted in that they are applicable only to interplanar ionic force constants, and we shall not discuss them. Our intention is to present only those transformations that are immediately applicable to a three-dimensional array of dipoles.

Finally, it should be pointed out that the Kellermann coefficients have certain symmetries with regard to interchanges of the various indices, for example,

$$^c\begin{bmatrix} 12 \\ xy \end{bmatrix} = {}^c\begin{bmatrix} 21 \\ xy \end{bmatrix} = {}^c\begin{bmatrix} 12 \\ yx \end{bmatrix} = {}^c\begin{bmatrix} 21 \\ yx \end{bmatrix}$$

and that by using Laplace's equation, one can establish the following condition:

$$\sum_\alpha {}^c\begin{bmatrix} kk' \\ \alpha\alpha \end{bmatrix} = 0$$

which provides a useful check on the derived coefficients.

Before concluding this section, we wish to present the derivation of what we term the "reduced coupling coefficients." These appear in problems to be discussed in the lattice-statics section of this volume, when we have to consider the electrostatic forces exerted by one ion in the crystal on the remaining ions. When this force array is subjected to a Fourier transformation, we obtain the reduced coupling coefficients, which we denote by $^c\begin{bmatrix} kk' \\ \alpha \end{bmatrix}$.

As before, we define

$$F^q(\mathbf{r}) = \sum_l f(\mathbf{r} - \mathbf{a}^l) \exp[2\pi i(\mathbf{q} \cdot \mathbf{a}^l)]$$

and

$$\frac{\partial}{\partial x}F^{\mathbf{q}}(\mathbf{r}) = \sum_l \frac{\partial}{\partial x}f(\mathbf{r} - \mathbf{a}^l)\exp[2\pi i(\mathbf{q}\cdot\mathbf{a}^l)]$$

$$= F^{\mathbf{q}}_x(\mathbf{r})$$

Then

$${}^c\!\begin{bmatrix} kk' \\ x \end{bmatrix} = e_k e_{k'} \lim_{\mathbf{r}\to-\mathbf{r}_{k'k}} \sum_l \frac{\partial}{\partial x}f(\mathbf{r} - \mathbf{a}^l)\exp\left\{2\pi i\left[\mathbf{q}\cdot\mathbf{r}\binom{l}{k'k}\right]\right\}$$

$$(k \neq k')$$

$$= e_k e_{k'} \lim_{\mathbf{r}\to-\mathbf{r}_{k'k}} \sum_l \frac{\partial}{\partial x}f(\mathbf{r} - \mathbf{a}^l)\exp\{2\pi i[\mathbf{q}\cdot(\mathbf{a}^l - \mathbf{r}_{kk'})]\}$$

Therefore

$${}^c\!\begin{bmatrix} kk' \\ x \end{bmatrix} = e_k e_{k'}\exp[-2\pi i(\mathbf{q}\cdot\mathbf{r}_{kk'})]F^{\mathbf{q}}_x(\mathbf{r}_{kk'}) \qquad (k \neq k')$$

Similarly

$${}^c\!\begin{bmatrix} kk \\ x \end{bmatrix} = e_k^2 \lim_{\mathbf{r}\to 0} \left\{\sum_l \frac{\partial}{\partial x}f(\mathbf{r} - \mathbf{a}^l)\exp[2\pi i(\mathbf{q}\cdot\mathbf{a}^l)] - \frac{\partial}{\partial x}f(\mathbf{r})\right\}$$

and therefore

$${}^c\!\begin{bmatrix} kk \\ x \end{bmatrix} = e_k^2 \lim_{\mathbf{r}\to 0} [F^{\mathbf{q}}_x(\mathbf{r}) - f_x(\mathbf{r})], \qquad f_x = \frac{\partial f}{\partial x}$$

Again

$$F^{\mathbf{q}}(\mathbf{r}) = (\pi v_a)^{-1} \sum_h |\mathbf{b}_h + \mathbf{q}|^{-2} \exp\left[-\frac{\pi^2}{E^2}|\mathbf{b}_h + \mathbf{q}|^2 + 2\pi i (\mathbf{b}_h + \mathbf{q}) \cdot (\mathbf{r}) \right]$$

$$+ \sum_l \frac{1 - G(E|\mathbf{r} - \mathbf{a}'|)}{|\mathbf{r} - \mathbf{a}'|} \exp[2\pi i(\mathbf{q} \cdot \mathbf{a}')]$$

where

$$G(s) = 2\pi^{-1/2} \int_0^s \exp(-\xi^2) \, d\xi$$

Now

$${}^c\!\begin{bmatrix} kk' \\ x \end{bmatrix} = e_k e_{k'} \exp[-2\pi i(\mathbf{q} \cdot \mathbf{r}_{kk'})] F_x^{\mathbf{q}}(\mathbf{r}_{kk'})$$

where

$$F^{\mathbf{q}}(\mathbf{r}_{kk'}) = (\pi v_a)^{-1} \sum_h |\mathbf{b}_h + \mathbf{q}|^{-2}$$

$$\times \exp\left[-\frac{\pi^2}{E^2}|\mathbf{b}_h + \mathbf{q}|^2 + 2\pi i(\mathbf{b}_h + \mathbf{q}) \cdot (\mathbf{r}_{kk'}) \right]$$

$$+ \sum_l \frac{1 - G(E|\mathbf{r}_{kk'} - \mathbf{a}'|)}{|\mathbf{r}_{kk'} - \mathbf{a}'|} \exp[2\pi i(\mathbf{q} \cdot \mathbf{a}')]$$

thus

$$\frac{\partial F^{\mathbf{q}}(\mathbf{r}_{kk'})}{\partial x_{kk'}} = \frac{2\pi i}{\pi v_a} \sum_h \frac{b_{hx} + q_x}{|\mathbf{b}_h + \mathbf{q}|^2} \exp\left[-\frac{\pi^2}{E^2}|\mathbf{b}_h + \mathbf{q}|^2 + 2\pi i(\mathbf{b}_h + \mathbf{q}) \cdot (\mathbf{r}_{kk'}) \right]$$

$$+ E \sum_l \frac{\partial \psi(E|\mathbf{r}_{kk'} - \mathbf{a}'|)}{\partial x_{kk'}} \exp[2\pi i(\mathbf{q} \cdot \mathbf{a}')]$$

Therefore

$$
{}^c\!\left[\begin{matrix} kk' \\ x \end{matrix}\right] = e_k e_{k'} \frac{2i}{v_a} \sum_h \frac{b_{hx} + q_x}{|\mathbf{b}_h + \mathbf{q}|^2} \exp\left[-\frac{\pi^2}{E^2}|\mathbf{b}_h + \mathbf{q}|^2 - 2\pi i (\mathbf{b}_h \cdot \mathbf{r}_{k'k}) \right]
$$

$$
+ E^2 \sum_l \psi'\!\left[E\left|\mathbf{r}\!\left(\begin{matrix} l \\ kk' \end{matrix}\right)\right|\right]
$$

$$
\times \left[x\!\left(\begin{matrix} l \\ kk' \end{matrix}\right)\Big/\left|\mathbf{r}\!\left(\begin{matrix} l \\ kk' \end{matrix}\right)\right|\right] \exp\left\{2\pi i\left[\mathbf{q}\cdot\mathbf{r}\!\left(\begin{matrix} l \\ k'k \end{matrix}\right)\right]\right\}\right\} \qquad (6.6)
$$

Similarly,

$$
{}^c\!\left[\begin{matrix} kk \\ x \end{matrix}\right] = e_k^2 \lim_{\mathbf{r}\to 0} [F_x^{\mathbf{q}}(\mathbf{r}) - f_x(\mathbf{r})]
$$

where

$$
F^{\mathbf{q}}(\mathbf{r}) = (\pi v_a)^{-1} \sum_h |\mathbf{b}_h + \mathbf{q}|^{-2} \exp\left[-\frac{\pi^2}{E^2}|\mathbf{b}_h + \mathbf{q}|^2 + 2\pi i (\mathbf{b}_h + \mathbf{q})\cdot(\mathbf{r}) \right]
$$

$$
+ \frac{1 - G(E|\mathbf{r}|)}{|\mathbf{r}|} + \sum_{l\neq 0} \frac{1 - G(E|\mathbf{r} - \mathbf{a}^l|)}{|\mathbf{r} - \mathbf{a}^l|} \exp[2\pi i (\mathbf{q}\cdot\mathbf{a}^l)]
$$

Now consider the term

$$
\frac{\partial}{\partial x}\left[\frac{1 - G(E|\mathbf{r}|)}{\mathbf{r}}\right] - \frac{\partial}{\partial x}\frac{1}{|\mathbf{r}|}
$$

$$
= -\frac{\partial}{\partial x}\frac{G(E|\mathbf{r}|)}{|\mathbf{r}|}
$$

$$
= -\frac{\partial}{\partial x}\frac{1}{|\mathbf{r}|}\left[\frac{2}{\pi^{1/2}}\int_0^{E|\mathbf{r}|} \exp(-\xi^2)\,d\xi\right]
$$

$$= -\frac{2}{\pi^{1/2}} \frac{\partial}{\partial x} \frac{1}{|\mathbf{r}|} \int_0^{E|\mathbf{r}|} \left(1 - \xi^2 + \frac{\xi^4}{2!} - \frac{\xi^6}{3!} + \cdots \right) d\xi$$

$$= -\frac{2}{\pi^{1/2}} \frac{\partial}{\partial x} \frac{1}{|\mathbf{r}|} \left(E|\mathbf{r}| - \frac{E^3|\mathbf{r}|^3}{3} + \frac{E^5|\mathbf{r}|^5}{10} - \cdots \right)$$

$$= -\frac{2}{\pi^{1/2}} \frac{\partial}{\partial x} \left(E - \frac{E^3|\mathbf{r}|^2}{3} + \frac{E^5|\mathbf{r}|^4}{10} - \cdots \right)$$

The limit of this is zero as $\mathbf{r}$ approaches zero. Now

$$\frac{\partial F^q(\mathbf{r})}{\partial x} = \frac{2\pi i}{\pi v_a} \sum_h \frac{b_{hx} + q_x}{|\mathbf{b}_h + \mathbf{q}|^2} \exp\left[-\frac{\pi^2}{E^2}|\mathbf{b}_h + \mathbf{q}|^2 + 2\pi i(\mathbf{b}_h + \mathbf{q}) \cdot (\mathbf{r}) \right]$$

$$+ \sum_l \frac{\partial}{\partial x} \frac{1 - G(E|\mathbf{r} - \mathbf{a}'|)}{|\mathbf{r} - \mathbf{a}'|} \exp[2\pi i(\mathbf{q} \cdot \mathbf{a}')]$$

Whence

$$\begin{bmatrix} kk \\ x \end{bmatrix} = e_k^2 \lim_{\mathbf{r} \to 0} \left\{ \frac{2i}{v_a} \sum_h \frac{b_{hx} + q_x}{|\mathbf{b}_h + \mathbf{q}|^2} \exp\left[-\frac{\pi^2}{E^2}|\mathbf{b}_h + \mathbf{q}|^2 \right. \right.$$

$$\left. \left. + 2\pi i(\mathbf{b}_h + \mathbf{q}) \cdot (\mathbf{r}) \right] \right\}$$

$$+ e_k^2 \lim_{\mathbf{r} \to 0} \left\{ \frac{\partial}{\partial x} \left[\frac{1 - G(E|\mathbf{r}|)}{|\mathbf{r}|} \right] - \frac{\partial}{\partial x} \frac{1}{|\mathbf{r}|} \right\}$$

$$+ e_k^2 \lim_{\mathbf{r} \to 0} \left\{ E \sum_{l \neq 0} \frac{\partial \psi(E|\mathbf{r} - \mathbf{a}'|)}{\partial x} \exp[2\pi i(\mathbf{q} \cdot \mathbf{a}')] \right\}$$

TABLE V. Reduced Coupling Coefficients $c\left[^{kk'}_{\alpha}\right]$ in Units of ie^2/r_0^2 for the NaCl-Structure Lattice

$2r_0 q_x$	$2r_0 q_y$	$2r_0 q_z$	$c\left[^{11}_x\right]$	$c\left[^{11}_y\right]$	$c\left[^{11}_z\right]$	$-c\left[^{12}_x\right]$	$-c\left[^{12}_y\right]$	$-c\left[^{12}_z\right]$
1.0	0.4	0.0	0.0000 0000	0.0275 2736	0.0000 0000	0.0000 0000	1.0768 9083	0.0000 0000
1.0	0.2	0.2	0.0000 0000	-0.0228 0170	-0.0228 0170	0.0000 0000	0.6957 9295	0.6957 9295
1.0	0.2	0.0	0.0000 0000	0.0447 2625	0.0000 0000	0.0000 0000	0.6682 2089	0.0000 0000
1.0	0.0	0.0	0.0000 0000	0.0000 0000	0.0000 0000	0.0000 0000	0.0000 0000	0.0000 0000
0.9	0.5	0.1	0.1574 9400	0.0000 0000	-0.1574 9400	0.4435 4435	1.1683 6540	0.4435 4435
0.9	0.3	0.3	0.2073 4222	-0.1421 1002	-0.1421 1002	0.4870 1233	1.0336 1012	1.0336 1012
0.9	0.3	0.1	0.2951 1286	0.0519 2985	-0.0364 8713	0.5236 8672	0.9483 9680	0.3879 2922
0.9	0.1	0.1	0.3912 3788	0.0322 9932	0.0322 9932	0.5876 6226	0.3642 3590	0.3642 3590
0.8	0.6	0.0	0.2143 3352	0.0941 4260	0.0000 0000	0.7893 5074	1.1398 7851	0.0000 0000
0.8	0.4	0.2	0.4003 3441	0.0447 2290	-0.1280 1307	0.9286 8282	1.2076 1672	0.8088 8408
0.8	0.4	0.0	0.4899 3011	0.1703 7863	0.0000 0000	0.9557 2864	1.1674 4681	0.0000 0000
0.8	0.2	0.2	0.6510 2916	0.0732 0891	0.0732 0891	1.0737 3752	0.7534 8165	0.7534 8165
0.8	0.2	0.0	0.7565 6390	0.1474 4980	0.0000 0000	1.1424 2704	0.7414 0484	0.0000 0000
0.8	0.0	0.0	0.8718 2639	0.0000 0000	0.0000 0000	1.2289 4925	0.0000 0000	0.0000 0000
0.7	0.7	0.1	0.1421 1002	0.1421 1002	-0.2073 4222	1.0336 1013	1.0336 1013	0.4870 1233
0.7	0.5	0.3	0.2667 4043	0.0000 0000	-0.2667 4043	1.2167 3741	1.3967 4543	1.2167 3741
0.7	0.5	0.1	0.5036 3758	0.2791 7821	-0.0712 8288	1.2354 0266	1.3289 1230	0.4503 3050
0.7	0.3	0.3	0.6706 5888	0.1021 8061	0.1021 8061	1.3844 7786	1.1406 2143	1.1406 2143
0.7	0.3	0.1	0.9520 7351	0.3251 2417	0.0643 1398	1.5415 6993	1.1363 9703	0.4424 4908
0.7	0.1	0.1	1.2934 9541	0.1520 2912	0.1520 2912	1.8026 5556	0.4536 4816	0.4536 4816
0.6	0.6	0.2	0.3170 8132	0.3170 8132	-0.1679 8331	1.3646 7438	1.3646 7438	0.8888 1156
0.6	0.6	0.0	0.4598 4126	0.4598 4126	0.0000 0000	1.3604 0575	1.3604 0575	0.0000 0000
0.6	0.4	0.4	0.4225 5907	0.0791 4851	0.0791 4851	1.5197 8048	1.4377 5186	1.4377 5186
0.6	0.4	0.2	0.8657 1364	0.4822 4164	0.1311 3244	1.6927 4983	1.4711 2672	0.9070 6080

TABLE V (contd.)

$2r_0q_x$	$2r_0q_y$	$2r_0q_z$	$c^{[11]}_x$	$c^{[11]}_y$	$c^{[11]}_z$	$-c^{[12]}_x$	$-c^{[12]}_v$	$-c^{[12]}_z$
0.6	0.4	0.0	1.0513 9864	0.6421 4593	0.0000 0000	1.7879 5593	1.5085 5712	0.0000 0000
0.6	0.2	0.2	1.4473 5440	0.4048 7078	0.4048 7078	2.1177 6940	0.9900 4375	0.9900 4375
0.6	0.2	0.0	1.7194 2271	0.5232 5614	0.0000 0000	2.3344 6808	1.0428 7073	0.0000 0000
0.6	0.0	0.0	2.0449 9065	0.0000 0000	0.0000 0000	2.6160 3799	0.0000 0000	0.0000 0000
0.5	0.5	0.5	0.0000 0000	0.0000 0000	0.0000 0000	1.5497 9519	1.5497 9519	1.5497 9519
0.5	0.5	0.3	0.5195 5989	0.5195 5989	0.1407 1851	1.6195 3998	1.6195 3998	1.2714 0352
0.5	0.5	0.1	0.8647 6147	0.8647 6147	0.0892 8071	1.7373 0345	1.7373 0345	0.4969 7728
0.5	0.3	0.3	1.1943 4195	0.6250 2072	0.6250 2072	2.0428 4698	1.4632 9385	1.4632 9385
0.5	0.3	0.1	1.7519 9560	1.0021 7167	0.3041 8214	2.4636 2566	1.6842 4179	0.6135 0963
0.5	0.1	0.1	2.5980 4439	0.4939 8962	0.4939 8962	3.2175 1427	0.7492 3881	0.7492 3881
0.4	0.4	0.4	0.6613 0054	0.6613 0054	0.6613 0054	1.6855 0158	1.6855 0158	1.6855 0158
0.4	0.4	0.2	1.2857 2628	1.2857 2628	0.5761 3341	2.0813 0874	2.0813 0874	1.1775 6726
0.4	0.4	0.0	1.5876 8863	1.5876 8863	0.0000 0000	2.3042 4921	2.3042 4921	0.0000 0000
0.4	0.2	0.2	2.4147 0348	1.1710 6975	1.1710 6975	3.0692 7662	1.6483 3629	1.6483 3629
0.4	0.2	0.0	3.0888 4542	1.5188 3293	0.0000 0000	3.6928 1900	1.9522 2993	0.0000 0000
0.4	0.0	0.0	4.0907 1155	0.0000 0000	0.0000 0000	4.6535 8561	0.0000 0000	0.0000 0000
0.3	0.3	0.3	1.4904 9869	1.4904 9869	1.4904 9869	2.1521 2625	2.1521 2625	2.1521 2625
0.3	0.3	0.1	2.4508 1506	2.4508 1506	0.7995 4251	3.0135 0641	3.0135 0641	1.0468 2501
0.3	0.1	0.1	4.7547 5951	1.5744 8539	1.5744 8539	5.2481 9358	3.2484 5197	1.7864 7037
0.2	0.2	0.2	2.8514 1347	2.8514 1347	2.8514 1347	3.2484 5197	3.2484 5197	3.2484 5197
0.2	0.2	0.0	4.5238 5365	4.5238 5365	0.0000 0000	4.8917 9096	4.8917 9096	0.0000 0000
0.2	0.0	0.0	9.5266 7105	0.0000 0000	0.0000 0000	9.8705 6515	0.0000 0000	0.0000 0000
0.1	0.1	0.1	6.4265 0468	6.4265 0468	6.4265 0468	6.6132 0479	6.6132 0479	6.6132 0479

Thus

$$
{}^{c}\begin{bmatrix} kk \\ x \end{bmatrix} = e_k^2 \left\{ \frac{2i}{v_a} \sum_h \frac{b_{hx} + q_x}{|\mathbf{b}_h + \mathbf{q}|^2} \exp\left(-\frac{\pi^2}{E^2} |\mathbf{b}_h + \mathbf{q}|^2 \right) \right.
$$

$$
\left. - E^2 \sum_{l \neq 0} \left[\psi'(E|\mathbf{a}^l|) \frac{a_x^l}{|\mathbf{a}^l|} \right] \exp[2\pi i (\mathbf{q} \cdot \mathbf{a}^l)] \right\} \qquad (6.7)
$$

We shall have occasion to pick out the leading term in the sum over reciprocal space in the lattice-statics part of this volume, where it is important to know the behavior of the reduced coupling coefficients for small wave vectors where this term is dominant.

The foregoing derivation can be extended to treat defect-lattice interactions for the case of an extraneous charge that is not located at a lattice site.

Again, for reference purposes we show in Table V a tabulation of these coefficients for the rock-salt structure for the same set of points as those in Table III.

IV

Theoretical and Experimental Single-Phonon Data

7. Comparison of Theoretical and Experimental Debye–Waller and Specific-Heat Data

From the dynamical matrix constructed according to the arguments in Section 5 we can calculate the normal-mode eigen-frequencies and also the eigenvectors; that is, the eigenvalues and eigenvectors of the matrix given by Eqs. (2.9) and (5.20). We employ periodic boundary conditions, which means that we are restricting ourselves to a crystal supercell whose symmetry is the same as that of the primitive unit cell, but whose sides are some many powers of 10 larger than those of the primitive unit cell. We impose periodic boundary conditions to the surfaces of the supercell. In practice, the calculations are done numerically by computer, which requires that we restrict ourselves to a finite sample of wave vectors. We thus have to decide how fine this sample has to be before the frequency distribution; that is, the function

$$N(\Omega) = \lim_{\substack{\Delta\Omega\to 0 \\ N\to\infty}} (6N\Delta\Omega)^{-1} \sum_{\mathbf{q}j} \int_{\Omega}^{\Omega+\Delta\Omega} \delta\left[\Omega' - \omega\left(\begin{matrix}\mathbf{q}\\j\end{matrix}\right)\right] d\Omega' \quad (7.1)$$

is an adequate representation of the true distribution function for an infinite crystal, since the enhancement of the sample density is equivalent to treating a larger and larger supercell [note that $\int_0^\infty N(\Omega)\, d\Omega = 1$].

We have devoted a great deal of computational time to examining this problem, while Gilat, Dolling, and Raubenheimer[106–108] have developed an interpolation procedure that provides a means of extracting many more frequencies than those calculated by directly diagonalizing the dynamical matrix. This technique is much faster than root-sampling procedures that attempt to obtain the effective infinite-crystal frequency distribution by increasing the number of sampled wave vectors. However, before any interpolation technique is used, its results should be checked against those of root-sampling calculations carried to a point that gives a good representation of the crystal frequency distribution.

As a test case some years ago,[109] we carried out a root-sampling calculation on sodium chloride. The density of the sample mesh of **q** vectors was increased from $10 \times 10 \times 10$ to $80 \times 80 \times 80$. At that time, this represented the practical limit of what could be done with an IBM 7094 computer. The programs were subsequently rewritten, and the amount of time required now is on the order of a few minutes on a CDC 7600 computer.

The calculated frequency spectrum was derived by a histogram technique: taking finite frequency intervals, counting the number of frequencies within each interval, and using these quantities to delimit blocks in a histogram. As the sample density is increased, the size of the histogram blocks can be reduced. The effect of this procedure was extremely dramatic. The frequency spectrum of sodium chloride, calculated from the deformation-dipole model, was extremely complicated. However, the corresponding frequency spectrum calculated from Kellermann's rigid-ion model was not appreciably different from that obtained by Kellermann himself in 1940.[23] This similarity seems to be fortuitous since the frequency spectrum of sodium chloride, within the rigid-ion approximation, has a particularly simple form and the subsidiary structure we obtained is not particularly marked. However, for the deformation-dipole model, the difference between the earlier spectrum calculated in 1959[72] and the later spectrum is extremely marked.

The results of our calculations made us reasonably confident that we had carried the root-sampling procedure to a point at which virtually all the features of the spectrum were clearly defined. More recently, we recalculated the frequency spectra of sodium chloride and certain other alkali halides by the interpolation procedure. In the case of sodium chloride, we could check the results obtained against the best earlier root-sampling calculations. The interpolation procedure considerably reduces (typically by a factor on the order of 3 or 4) the time needed to compute the frequency spectrum with the same degree of precision. However, the coarse mesh that must be used in the interpolation procedure to provide reliable results turns out to be reasonably dense ($30 \times 30 \times 30$).

Our calculations have revealed that the original idea of Van Hove[110] and, more recently, Phillips[111] that the frequency spectra of crystal lattices are dominated by the singularities that occur when

$$\left| \nabla_q \left[\omega \binom{q}{j} \right] \right| = 0 \qquad (7.2)$$

is not adequate for the alkali halides. The probable reason is the multitude of these singularities. One can attempt, using computed dispersion curves, to associate critical points in the frequency spectrum with points of vanishing slope on the dispersion curves, as we did in our 1966 paper.[109] However, it appears that there is no easy way of estimating the strength of a given singularity. In our experience, any modification of the dynamical matrix that shifts the frequencies of the optical branches with respect to the frequencies of the acoustic branches has an extremely complex effect on the frequency spectrum (cf. the rigid-ion model and the deformation-dipole model calculations for sodium chloride).

To facilitate the assignment of specific singularities to specific features of the dispersion curves, we have, in the past, presented figures showing both the frequency spectrum and the dispersion curves plotted with a common frequency scale. In Figs. 5–8 we show such plots for the four sodium halides.

Figures 5–8 show behavior that is broadly characteristic of the alkali halides, although it is not possible to make valid generalizations as to the detailed form of frequency spectrum to be expected for any given crystal. This should be evident from Figs. 5–8. However, one can see that, as the disparity between the masses of the two ions becomes greater, a clear separation between the optical and acoustic parts of the frequency spectrum develops. For sodium bromide and sodium iodide there is actually a gap in the frequency spectrum between the optical and acoustic branches. Nonetheless, as we shall see shortly, it is unwise to assume that the low-frequency end of the spectrum is entirely "acoustic" in character and that the

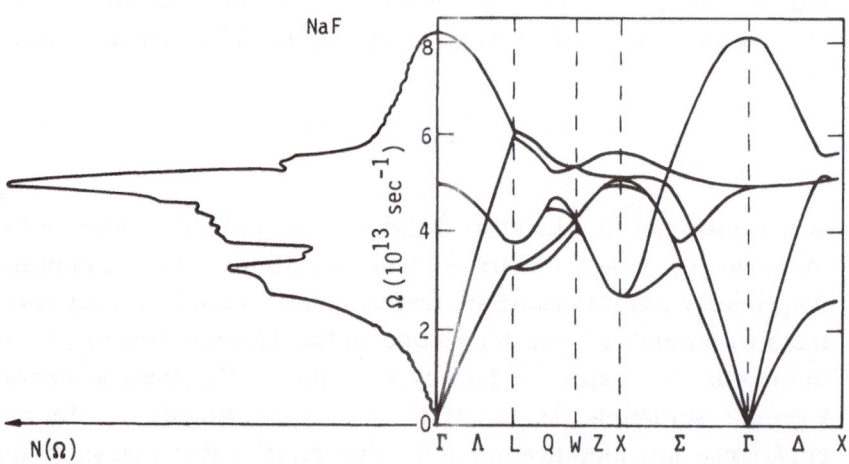

FIG. 5. Dispersion curves and frequency spectrum for NaF. The model used is the deformation-dipole model with short-range forces extended to include closest-neighbor negative ions. The calculation has been carried out for low-temperature input data.

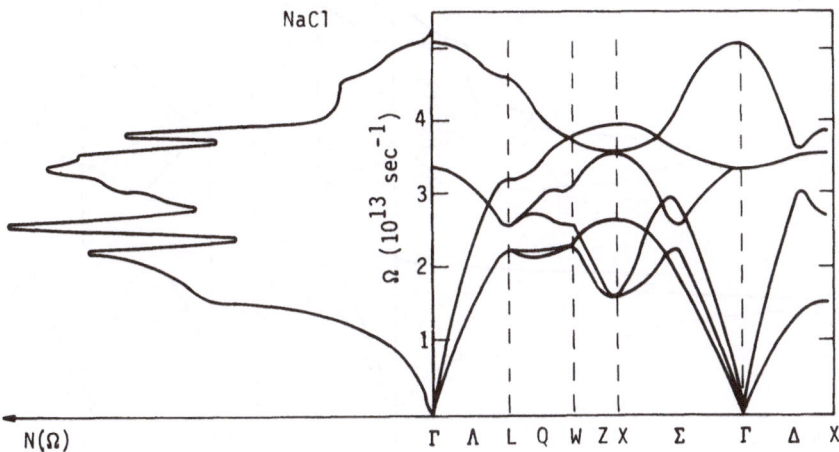

FIG. 6. Dispersion curves and frequency spectrum for NaCl. The model used is the deformation-dipole model with short-range forces extended to include closest-neighbor negative ions. The calculation has been carried out for low-temperature input data.

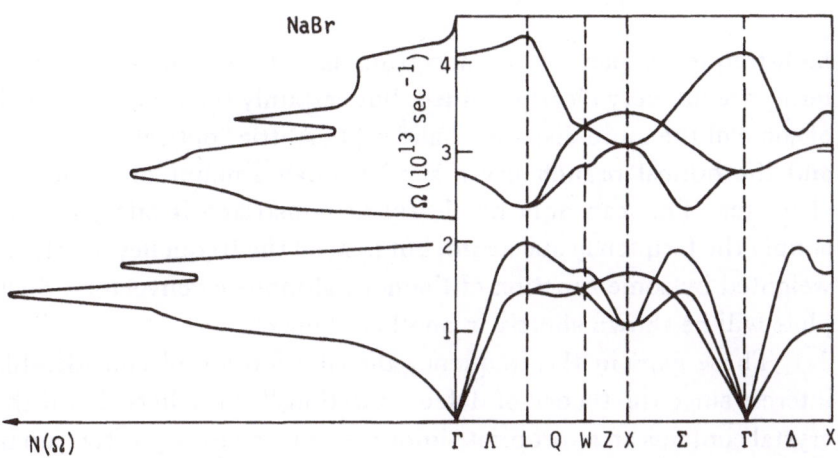

FIG. 7. Dispersion curves and frequency spectrum for NaBr. The model used is the deformation-dipole model with short-range forces extended to include closest-neighbor negative ions. The calculation has been carried out for low-temperature input data.

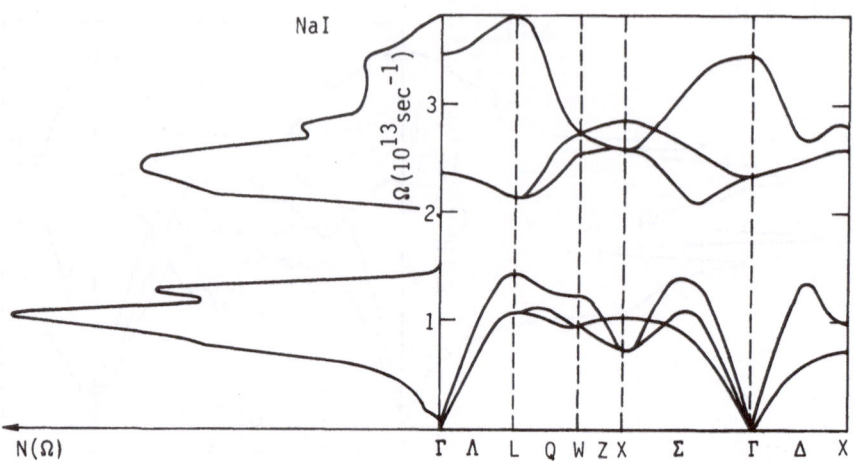

FIG. 8. Dispersion curves and frequency spectrum for NaI. The model used is the deformation-dipole model with short-range forces extended to include closest-neighbor negative ions. The calculation has been carried out for low-temperature input data.

high-frequency part of the spectrum is entirely "optical." These terms are not very clearly defined, but certainly the acoustic branch of some of the more disparate halides has a little "optical" character, and the optical branch has a similar small amount of "acoustic" character. This can only be clearly demonstrated if one plots not simply the frequency spectrum, but instead the frequency spectrum weighted by some function of the normal-mode eigenvectors. Such plots will be shown shortly in another context.

These gaps in the frequency distribution are of considerable interest since the theory of defect vibrations[112] predicts that if the crystal contains an appropriate impurity (e.g., a heavy positive ion in a halide whose positive ion is very much lighter than its negative ion), the impurity may produce a localized vibration whose frequency lies within the gap. This type of localized mode is often referred to as a "gap mode."

It is interesting to examine the computed frequency distributions and to observe the strengths of the various critical-point singularities. As already stated, there is no easy way of predicting the strength of any given singularity. Some of the singularities are extremely sharp and narrow. Moreover, there are singularities at which the distribution function drops almost vertically. These are interesting features when one is considering defect vibrations because these singularities are likely to be associated with resonances in the amplitude of the vibrating defect.

7.1. Debye–Waller Factors

We now have to consider the various derived properties of the frequency distribution, which itself is normally not measured directly. Some progress can be made in this direction by making use of incoherent inelastic neutron scattering[113] since many of the alkali halides have quite high incoherent-scattering cross sections. However, this will not determine the frequency distribution directly. Ideally, if one had a crystal in which the incoherent scattering came entirely from one type of atom, it would be possible to compute the following functions:

$$N_k(\Omega) = \lim_{\substack{\Delta\Omega \to 0 \\ N \to \infty}} (6N\Delta\Omega)^{-1} \sum_{\mathbf{q}j} \int_{\Omega}^{\Omega + \Delta\Omega} \delta\left[\Omega' - \omega\left(\begin{matrix}\mathbf{q}\\j\end{matrix}\right)\right] \left|\mathbf{e}\left(k\left|\begin{matrix}\mathbf{q}\\j\end{matrix}\right.\right)\right|^2 d\Omega'$$

$$(7.3)$$

This weighted density of states is also something that can be calculated, and it is of interest to do so since one can assess to some extent the degree to which a particular part of the spectrum is optical or acoustic. More precisely, one can scan the frequency spectrum to see where the amplitudes of the vibrations are predominantly those of the lighter atom and where they are those of the heavier atom. Figures 9–12 show the weighted densities of states computed

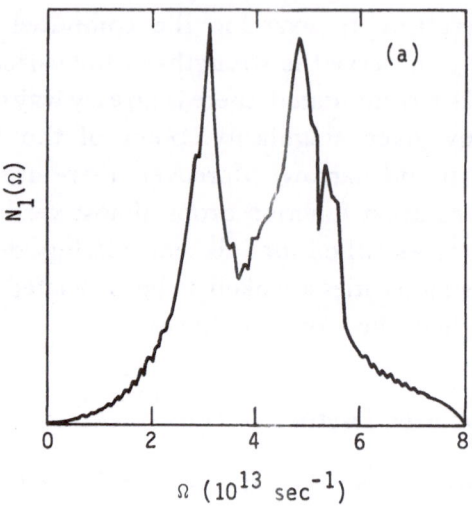

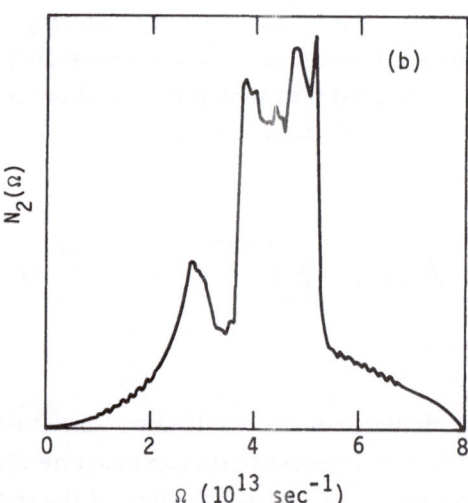

FIG. 9. Weighted densities of states for NaF computed according to Eq. (7.3): (a) $k = 1$ (positive ion); (b) $k = 2$ (negative ion). The model used is the deformation-dipole model with short-range forces extended to include closest-neighbor negative ions. The calculation has been carried out for low-temperature input data.

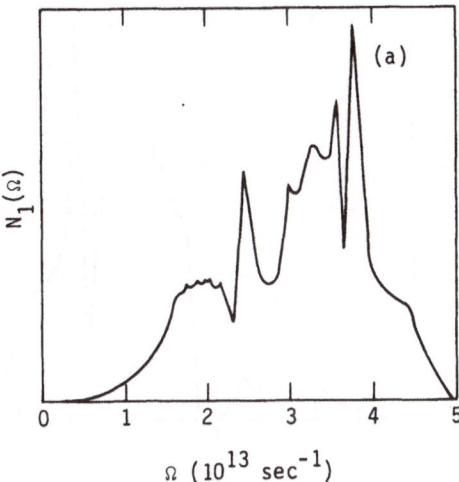

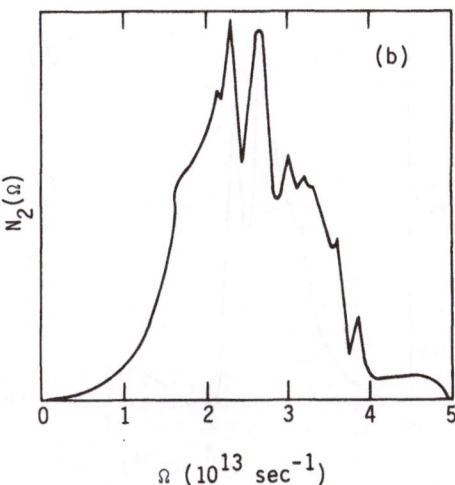

FIG. 10. Weighted densities of states for NaCl computed according to Eq. (7.3): (a) $k = 1$ (positive ion); (b) $k = 2$ (negative ion). The model used is the deformation-dipole model with short-range forces extended to include closest-neighbor negative ions. The calculation has been carried out for low-temperature input data.

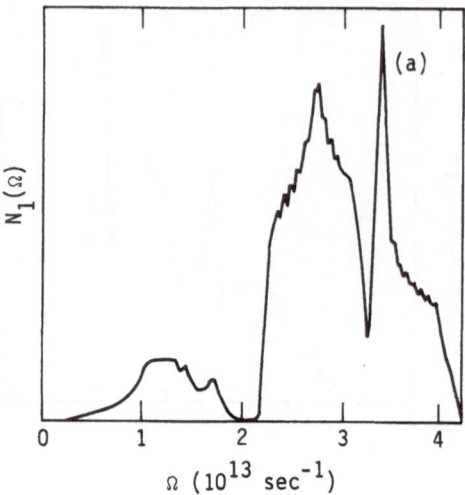

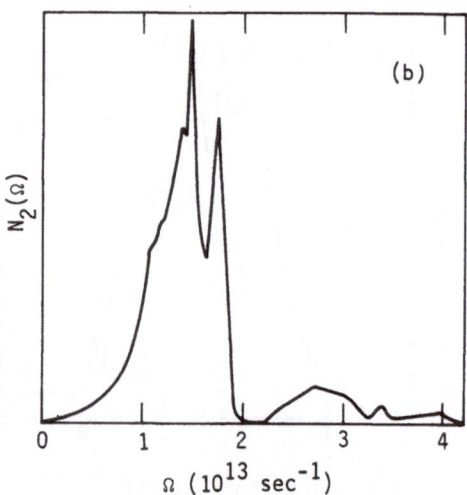

FIG. 11. Weighted densities of states for NaBr computed according to Eq. (7.3): (a) $k = 1$ (positive ion); (b) $k = 2$ (negative ion). The model used is the deformation-dipole model with short-range forces extended to include closest-neighbor negative ions. The calculation has been carried out for low-temperature input data.

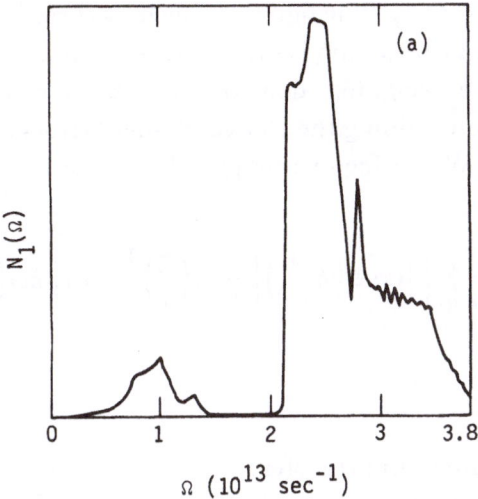

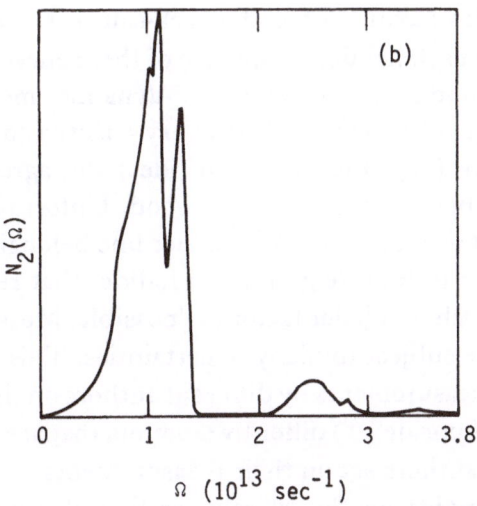

FIG. 12. Weighted densities of states for NaI computed according to Eq. (7.3): (a) $k = 1$ (positive ion); (b) $k = 2$ (negative ion). The model used is the deformation-dipole model with short-range forces extended to include closest-neighbor negative ions. The calculation has been carried out for low-temperature input data.

according to Eq. (7.3). The figures show how these functions vary as the mass of the negative ion is increased from that of fluorine to that of iodine. The weighted densities of states can be indirectly measured by determining the Debye–Waller factors. The expression for the Debye–Waller factor is $\exp(-2W_k)$, where

$$
2W_k = \frac{\hbar}{2Nm_k} \sum_{\mathbf{q}j} \left[\left| \boldsymbol{\kappa} \cdot \mathbf{e}\left(k \left| \begin{matrix} \mathbf{q} \\ j \end{matrix} \right. \right) \right|^2 \middle/ \omega\left(\begin{matrix} \mathbf{q} \\ j \end{matrix} \right) \right] \coth\left[\hbar\omega\left(\begin{matrix} \mathbf{q} \\ j \end{matrix} \right) \middle/ 2k_B T \right]
$$

$$
= \frac{1}{8\pi^2} B_k \kappa^2 \tag{7.4}
$$

$\boldsymbol{\kappa}$ being the momentum transfer.

Calculations of this type have been made by Jaswal[114] and ourselves,[115] and we show in Figs. 13–18 the behavior of the B parameters in Eq. (7.4) as a function of temperature T for a number of alkali halides having either the rock-salt or the cesium chloride structure. For each halide, a sequence of these curves is shown, each curve corresponding to a given lattice-dynamics model.

The curves obtained are clearly very sensitive to the model used. In many cases (e.g., the cesium halides) the agreement between theory and experiment is extremely good. Unfortunately, it is only for those crystals that have a Mössbauer line belonging to one of the constituents or to both (e.g., cesium iodide) that reliable measurements of the Debye–Waller factor are possible. Measurements made with x rays are subject to many uncertainties. This is illustrated by the fact that measurements by different authors on the same material (e.g., sodium fluoride[116]) differ by amounts that are well outside the variances the authors set on their measurements.

Measurements on the Mössbauer line of some appropriately chosen substitutional impurity present a much more complicated situation. It has been shown[117] that the modification of the perfect-lattice normal modes by an impurity drastically affects the Debye–Waller factor. Furthermore, there is no easy way of calculating this effect without knowing in detail the nature of the force-constant changes produced by the impurity. This current lack of precision precludes the use of such measurements as the basis for precise statements about the eigenvectors and eigenfrequencies of the host crystal.

However, when direct measurements of the Debye–Waller factor can be made on pure materials, they are of interest because the Debye–Waller factor depends on the frequency spectrum of the crystal in such a way that it is sensitive to the shape of the whole spectrum. At high temperatures the Debye–Waller factor is proportional to the inverse second moment of the weighted frequency distribution [see Eq. (7.4)] when we use the fact that

$$\frac{1}{2}\coth\left[\hbar\omega\binom{\mathbf{q}}{j}\Big/2k_{\mathrm{B}}T\right] \to k_{\mathrm{B}}T\Big/\hbar\omega\binom{\mathbf{q}}{j}$$

as T approaches infinity, whereas at low temperatures it is proportional to the inverse first moment of this distribution. In the intermediate range, no such simple relationship applies. On balance, the Debye–Waller factor contains more information about the frequency distribution and eigenvectors than the specific-heat measurements, which we shall discuss shortly. In particular, specific heats give us no information regarding the eigenvectors. The type of information that can be extracted from the specific heats has been discussed by Barron and co-workers,[118] who have shown that it is possible to correlate the Debye–Waller factor and the specific heat. However, it is not clear to us how they can do this without using information about the eigenvectors that is not obtainable from the specific-heat data alone.

The temperature dependence of the calculated Debye–Waller B parameters often displays interesting features. Cesium bromide, for example, shows rather surprising behavior in that for two of the crystal models, DD and DD′, the computed B values for the two

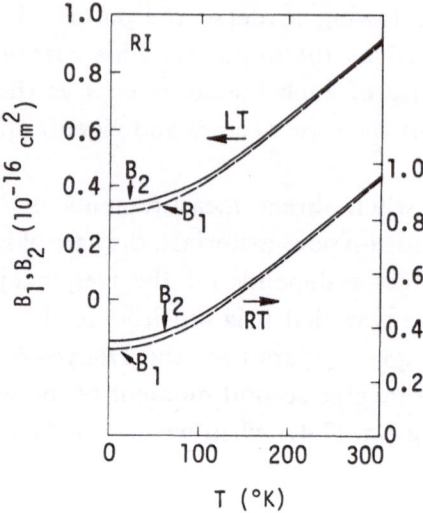

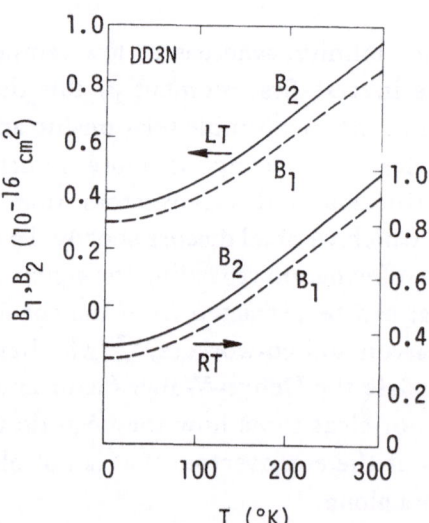

FIG. 13. Debye–Waller parameters B_k for NaF computed according to Eq. (7.4) for the following models: RI (rigid ion); DD (deformation dipole with first-neighbor repulsive forces only); DD3N (deformation dipole with short-range forces extended to include closest-neighbor negative ions); and DD3N(AB) (an extension of the

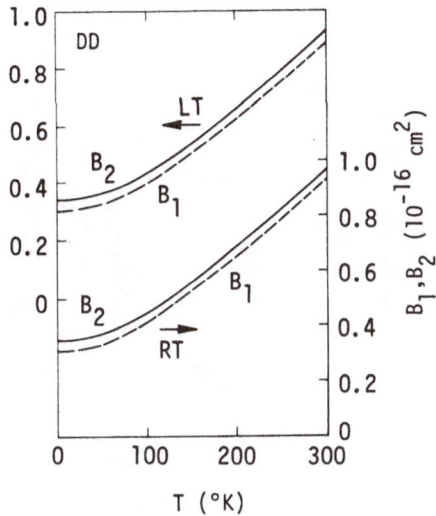

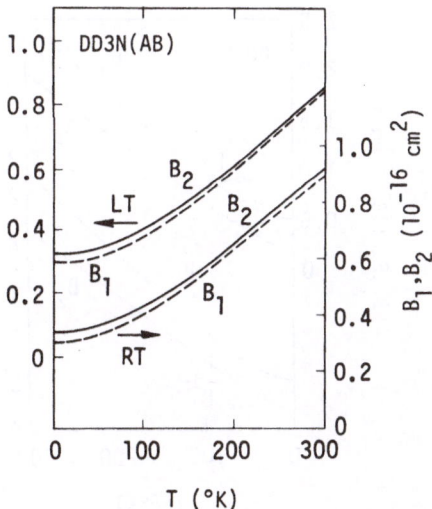

DD3N model in which force constants that resist change of the 90° angles between nearest-neighbor bonds are included). The LT and RT curves correspond to low- and room-temperature input data respectively; the scales for the curves are given at the left or right side of the graph as indicated by the arrows.

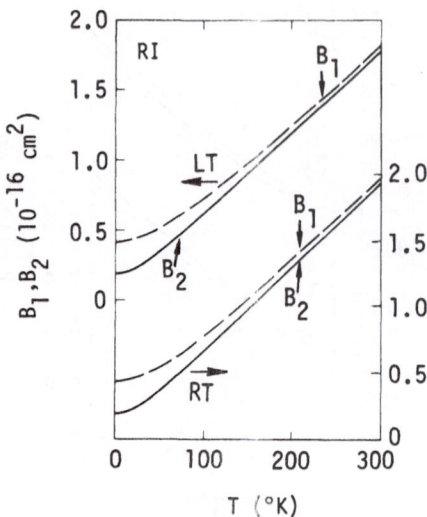

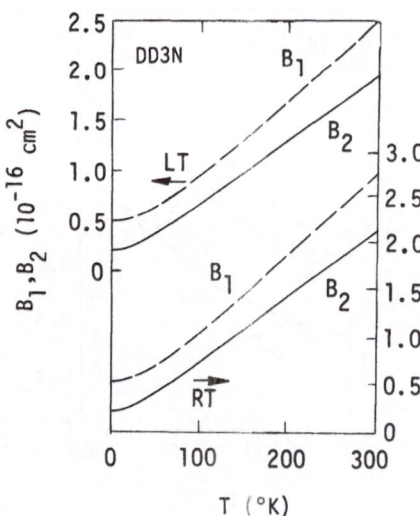

FIG. 14. Debye–Waller parameters B_k for NaI computed as in Fig. 13.

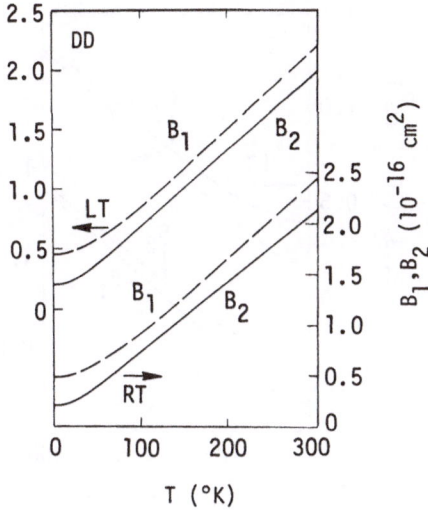

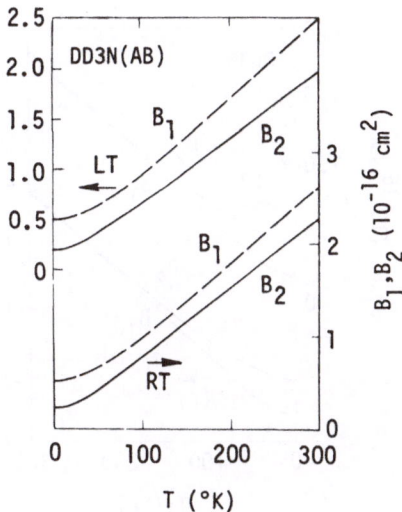

FIG. 14 (*contd.*)

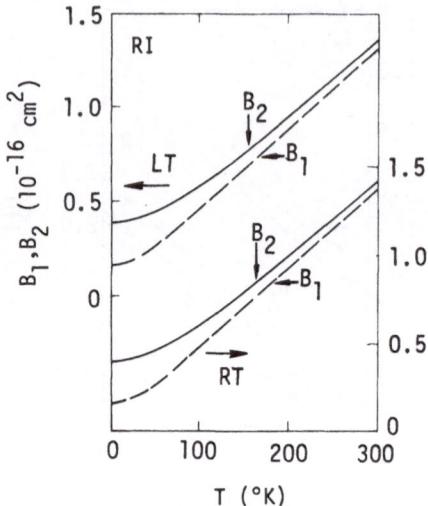

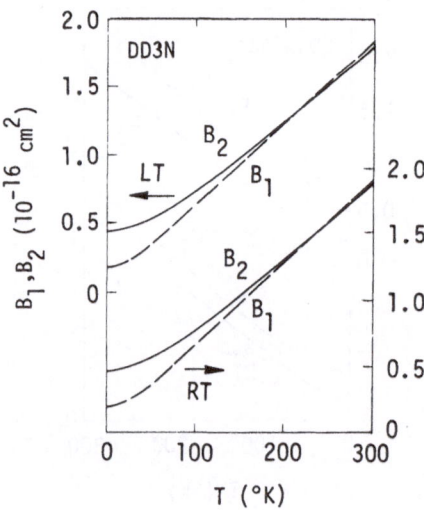

FIG. 15. Debye–Waller parameters B_k for CsF computed as in Fig. 13.

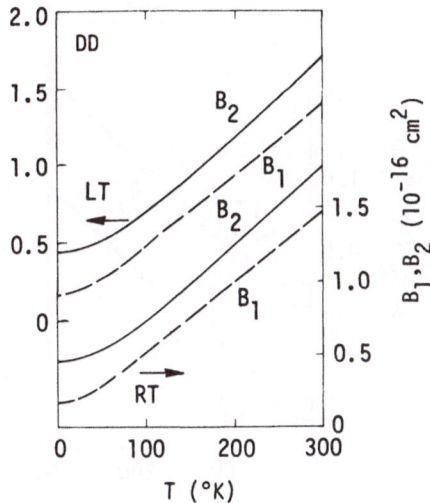

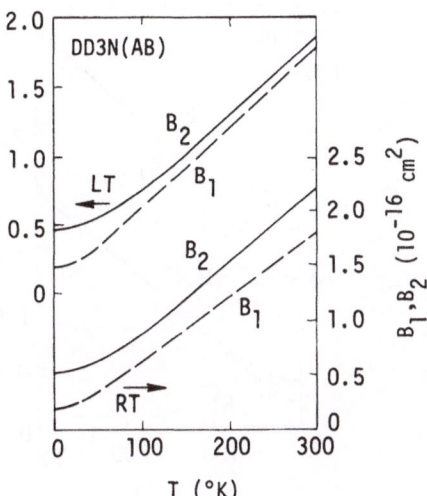

FIG. 15 (*contd.*)

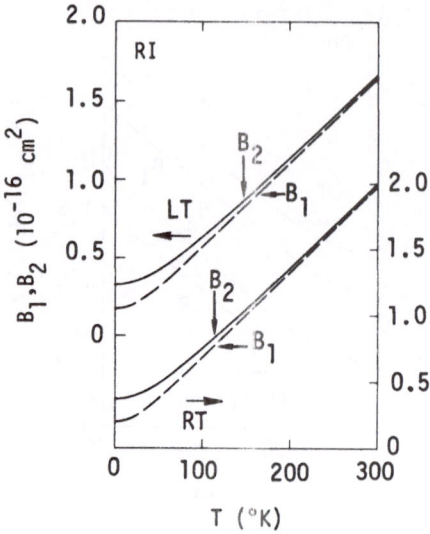

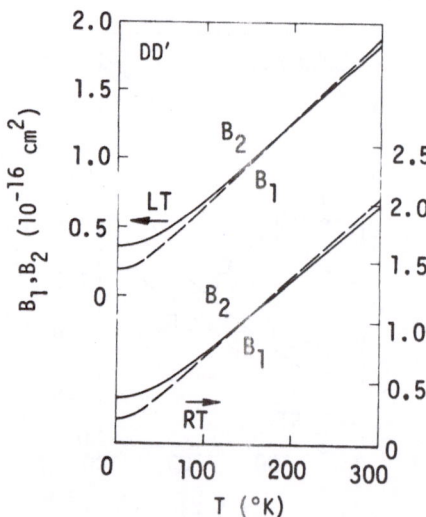

FIG. 16. Debye–Waller parameters B_k for CsCl computed according to Eq. (7.4) for the following models: RI (rigid ion); DD (deformation dipole with first-neighbor repulsive forces only); DD′ [variation of the DD model in which the compressibility used is derived from Eq. (4.26a), i.e., the first Szigeti relation]; and DD3N

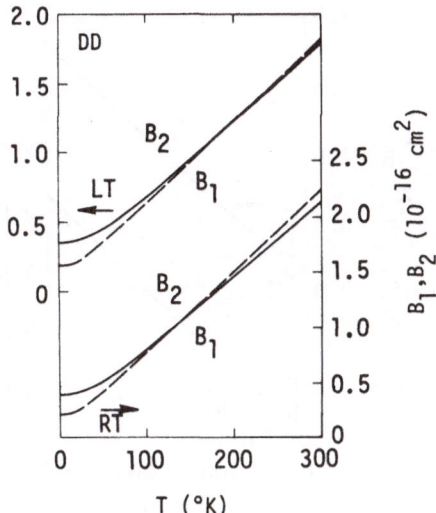

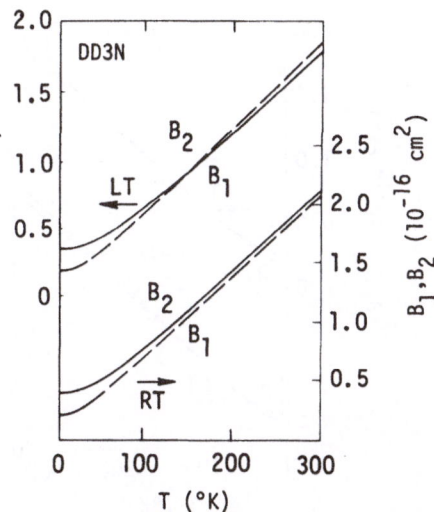

(deformation dipole with short-range forces extended to include closest-neighbor negative ions). The LT and RT curves correspond to low-temperature and room-temperature input data respectively; the scales for the curves are given at the left or right side of the graph as indicated by the arrows.

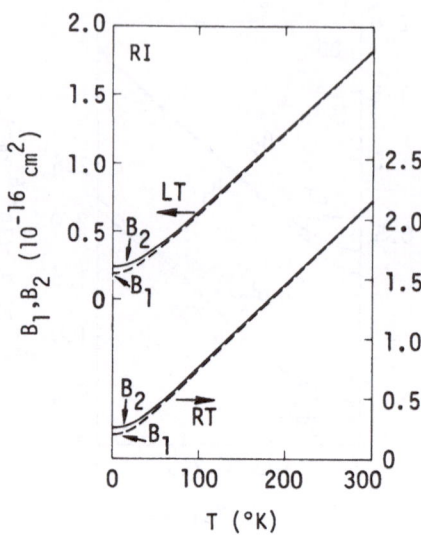

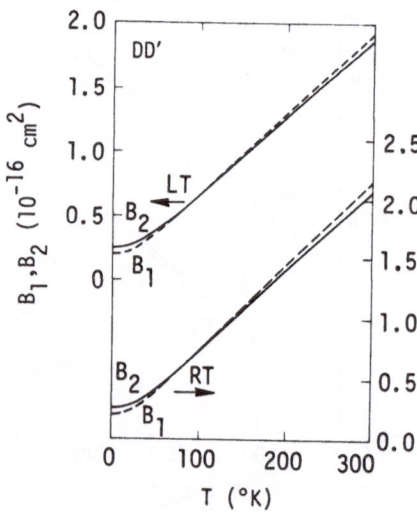

FIG. 17. Debye–Waller parameters B_k For CsBr computed as in Fig. 16.

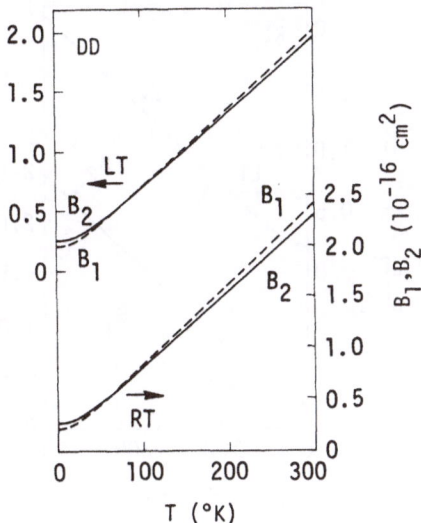

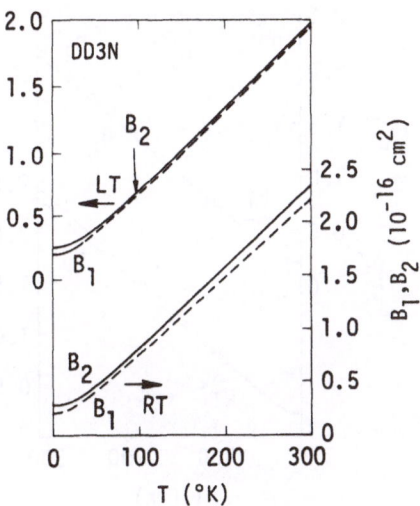

FIG. 17 (*contd.*)

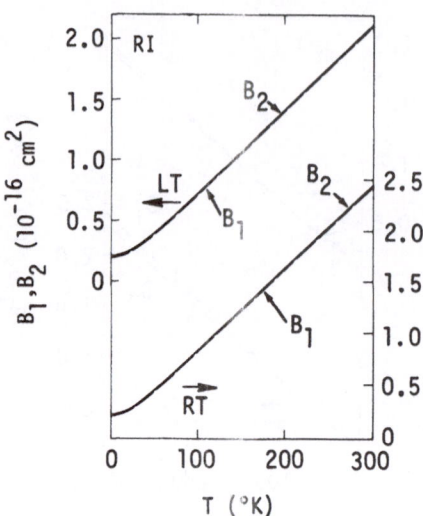

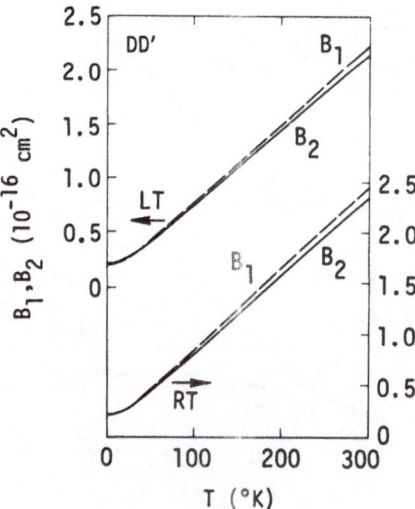

FIG. 18. Debye–Waller parameters B_k for CsI computed as in Fig. 16.

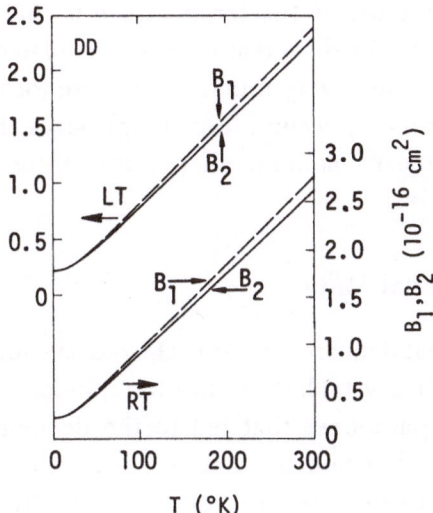

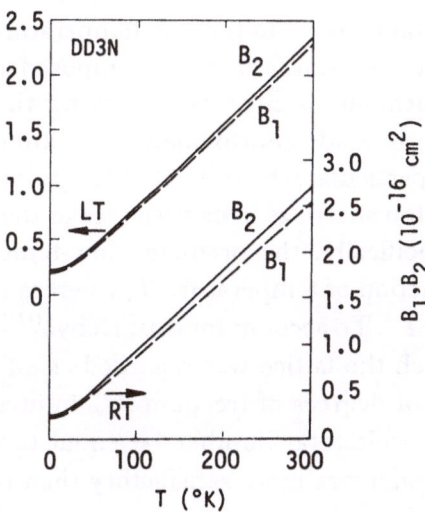

FIG. 18 (*contd.*)

constituent ions cross as the temperature is raised. This is rather intriguing because the dispersion curves calculated for these models do not appear to differ very much from those for the DD3N model, for which no crossing occurs. This emphasizes that the dispersion curves by no means contain all possible information about the dynamical matrix.

7.2. Specific-Heat Data

Specific-heat data are probably the oldest source of information about phonons in crystal lattices and were, in fact, one of the sources of the original paradoxes that led to the development of modern quantum theory. The nonconstancy of the specific heats of crystals as a function of temperature was not explicable by classical equipartition theory. However, when Einstein,[119] following Planck,[120] introduced the quantization of harmonic-oscillator energy levels, it became possible to account for the manner in which the specific heat of any real crystal behaves in the low-temperature limit.

In Einstein's model, the crystal is composed of a large number of independent harmonic oscillators all having the same frequency. This is a relatively crude approximation and thus subject to certain limitations. In particular, the predicted low-temperature rise in the specific heat is too sharp, as is its turnover to the high-temperature asymptote. Specifically, the predicted low-temperature rise is an exponential function of temperature T, whereas the observed rise is proportional to T^3. To account for this, Debye[121] in 1912 developed a model in which the lattice was regarded as an elastic continuum whose number of degrees of freedom was limited by imposing the constraint that their total number be equal to that of the whole crystal. This model was more satisfactory than the Einstein model since it did predict a T^3 dependence for the specific heat at very low temperatures.

In a sense, the Debye model did something of a disservice to solid-state theory as applied to lattice dynamics. At the time it was developed, and for many years thereafter, it seemed so satisfactory that it was regarded as the last word. There was also the added attraction that the Debye spectrum and functions derived therefrom had relatively simple analytic forms, and thus calculations of many crystal properties could be made without resort to extensive numerical computations. However, Debye himself almost certainly regarded his model as no more than a first approximation.[122] It was not until the mid-1930s that Blackman produced a series of papers[123–127] revealing the limitations of the Debye model and showing that they were much more serious than had been realized. Blackman's work involved a series of calculations of the specific heats of certain model lattices (e.g., a face-centered-cubic lattice with nearest-neighbor interactions) as a function of temperature. These were carried out using the correct lattice-dynamical equations to compute the normal-mode frequencies. Thence the specific heat was computed according to the equation

$$C_v = k_B \sum_{\mathbf{q}j} \left[\hbar\omega\binom{\mathbf{q}}{j} \middle/ k_B T \right]^2 \exp\left[\hbar\omega\binom{\mathbf{q}}{j} \middle/ k_B T \right]$$

$$\times \left\{ \exp\left[\hbar\omega\binom{\mathbf{q}}{j} \middle/ k_B T \right] - 1 \right\}^{-2} \tag{7.5}$$

For each of the models Blackman calculated an "effective" Debye temperature θ_D. This was done by taking the calculated specific heat at any given temperature and determining the appropriate Debye temperature that would give the same specific heat when used in a Debye function. In this way the effective Debye temperature could be determined as a function of temperature.

These results are revealing; in particular, they all show a minimum at temperatures such that $T/\theta_D \approx 0.1$. The presence of this minimum, provided it occurs at a reasonably low temperature, means that the observed specific heat will follow a T^3 law over an appreciable temperature range in the vicinity of the minimum.

However, this does not signal reaching the true Debye limit, at which the only appreciably excited set of lattice oscillations is the set of long-wavelength acoustic modes. This regime is reached only at very much lower temperatures ($T < \theta_D/50$). There has been controversy as to whether or not the limiting Debye temperature as T approaches zero is the same as that computed from the ultrasonically measured elastic constants.[128] In fact, the two do indeed correspond, but one has to be careful to determine the thermal value from the true low-temperature regime. If this is not done correctly, a discrepancy will appear because the curvature of the plots of the effective Debye temperature versus the temperature at absolute zero is generally quite sharp.

Blackman's work was not extensively followed up until much later. In the absence of modern high-speed digital computers, only very simple models of crystals were studied. Nonetheless, Kellermann,[129] in his classic work on sodium chloride, did make a Blackman type of calculation of the effective Debye temperature as a function of temperature. Furthermore, calorimetric techniques were such that it was not possible to obtain any but the crudest information about crystal frequency spectra by specific-heat measurements. It was not until the mid-1950s that Morrison and his group[130] developed precise measurement techniques and obtained very much more meaningful information about the frequency spectra from specific-heat data.

The amount of information that can be extracted is still limited, but low-temperature measurements that define the effective Debye temperature as a function of temperature are still extremely valuable. In particular, it is of great importance to examine in detail the shape of the minimum in this curve. Barron, Berg, and Morrison[131] showed that it is possible to extract information of the following form:

1. At low temperatures the specific heat is given by

$$C_v = aT^3 + bT^5 + cT^7 \qquad (7.6)$$

and Barron *et al.* were able to extract from their data the
coefficients a, b, and c.

2. At high temperatures the effective Debye temperature θ_∞ is
    related to the second moment of the frequency distribution
    as follows:

$$\theta_\infty = (\hbar/k_B)(5\mu_2/3)^{1/2} \tag{7.7}$$

where, in general,

$$\mu_n = \int_0^{\Omega_m} \Omega^n N(\Omega)\, d\Omega$$

and Ω_m is the high-frequency cutoff.

Moreover, the curvature away from the high-temperature
asymptote as the temperature is reduced makes it possible to extract
the fourth and sixth moments using the following expansion:

$$\theta_D^2 = \theta_\infty^2 \left[1 - A\left(\frac{\theta_\infty}{T}\right)^2 + B\left(\frac{\theta_\infty}{T}\right)^4 - \cdots \right] \tag{7.8}$$

where

$$A = \frac{3}{100}\left(\frac{\mu_4}{\mu_2^2} - \frac{25}{21}\right)$$

and

$$B = \frac{1}{1400}\left[\left(\frac{\mu_6}{\mu_2^3} - \frac{125}{81}\right) - 100A\right]$$

Then, for the intermediate region, it is possible to extract the various inverse moments $\mu_{-2.5}$, μ_{-2}, $\mu_{-1.5}$, and μ_{-1} by using

$$\frac{1}{6Nk_B} \int_0^\tau \frac{C_v^0}{T^n} dT = \Gamma(n+1)\zeta(n)\left(\frac{\hbar}{k_B}\right)^{1-n} \mu_{1-n} = \frac{1}{n-1}\frac{1}{\tau^{n-1}}$$

$$+ \sum_{s=1}^\infty (-1)^{s+1} \frac{B_{2s}}{(2s)!} \frac{2s-1}{2s+n-1}\left(\frac{\hbar}{k_B}\right)^{2s}$$

$$\times \frac{\mu_{2s}}{\tau^{2s+n-1}} \qquad (1 < n < 4) \qquad (7.9)$$

The B terms are the Bernoulli numbers, $\Gamma(n+1)$ and $\zeta(n)$ are the gamma and zeta functions, and C_v^0 is the "harmonic" specific heat.[131]

All these are directly calculable quantities for any particular model. At present we have not been able to find any satisfactory way of computing the coefficients in the low-temperature specific-heat expansion. However, the various moments and the θ_D curve are directly calculable. These moments and the θ_D curves, though interrelated, are best regarded as independent pieces of information. This is so because each is sensitive to a different region of the frequency spectrum. Specifically, it appears that the position of the minimum of the θ_D curve is largely influenced by the frequency at which the first Van Hove singularity in the frequency distribution occurs. Thus any computed spectrum that misplaces this first peak is likely to give a θ_D curve that seems to be in very bad agreement with experiment even though the rest of the frequency spectrum may, in fact, be quite good.

Thus we have the following single-phonon data that can be derived from experimental specific-heat measurements and should therefore be computed theoretically: we have the various moments, the θ_D curve, and also the specific heat itself. Although the specific heat is quite sensitive to the form of the frequency distribution, this dependence is not very well brought out by an actual examination of the specific-heat curve itself. It is only when the specific-heat function is converted to an effective θ_D that this sensitivity becomes apparent.

Before one can make a direct comparison it is necessary to apply various corrections to the experimental data in order to eliminate anharmonic effects. Obviously, one makes the normal correction in going from C_p to C_v:

$$C_p(T) - C_v(T) = TV\alpha^2/\varepsilon \qquad (7.10)$$

where α is the volume expansion coefficient, ε is the isothermal compressibility, and V is the specimen's volume.

There is then a further correction to be applied to allow for the fact that what we are actually calculating is the specific heat at some fixed volume for all temperatures, not the specific heat at constant volume at a given temperature. This correction is most significant in the use of the high-temperature expansion for θ_D, and Barron, Berg, and Morrison[131] made the following correction:

$$\frac{\theta_D(V_0)}{\theta_D(V_T)} = \left(\frac{\rho_0}{\rho_T}\right)^\gamma$$

where ρ_0 (ρ_T) and V_0 (V_T) are the density and volume at 0°K (T°K) and γ is the Grüneisen constant.

At the highest temperatures the effects of anharmonicity are such that they cannot be eliminated by simple types of correction. There are anharmonic corrections to the free energy and thus to the specific heat that arise even if the crystal volume is maintained completely fixed, and these have the effect of causing the computed θ_D curves to deviate from the ideal limiting value at high temperatures.

The deviations found by Morrison and his group [130] for the alkali halides with the rock-salt structure are quite different from

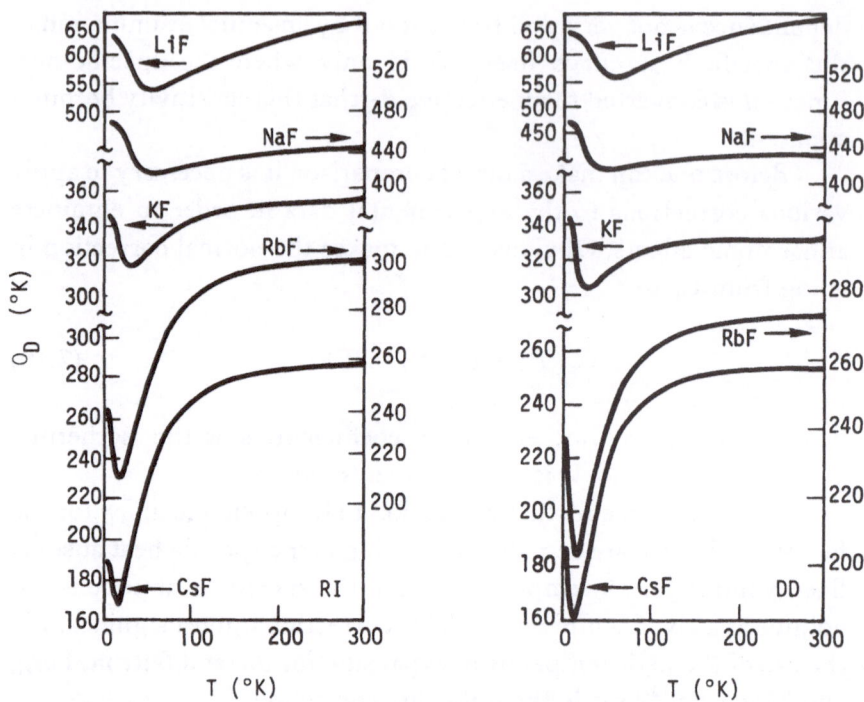

FIG. 19. Effective Debye temperature θ_D for the alkali fluorides computed for the following models: RI (rigid ion); DD (deformation dipole with first-neighbor repulsive forces only); DD3N (deformation dipole with short-range forces extended

those observed more recently by Sorai[132] for the cesium halides. This behavior indicates that the relative magnitudes of the third- and fourth-order anharmonic terms in the potential energy are very different in the cesium halide sequence, but such considerations lie outside the scope of this volume.

To update the types of comparison made in our earlier papers,[74,75] we present here a new set of calculated thermal data for the sequence of alkali halides. These results are shown in Figs. 19–28. Specifically, we present θ_D curves (Figs. 19–23) and moment functions (Figs. 24–28); the moments themselves are given in Table

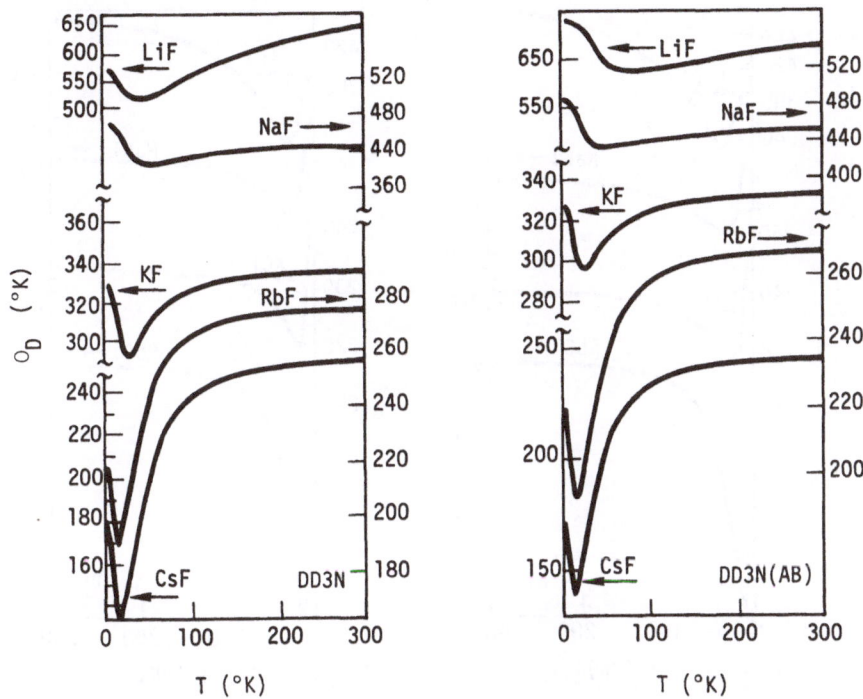

to include closest-neighbor negative ions); and DD3N(AB) (an extension of the DD3N model in which force constants that resist change of the 90° angles between nearest-neighbor bonds are included). Low-temperature input data have been used.

VI. For convenience we use the moment function

$$\omega_D(n) = [\tfrac{1}{3}(n + 3)\mu_n]^{1/n} \qquad (7.11)$$

introduced by Barron and Morrison.[131] This enables us to display the moment data on a common scale. For a true Debye spectrum, this moment function would be a constant. As can be seen, this is definitely not the case. Another point of interest is the very different shape of this function for alkali halides having the cesium chloride structure as compared with those having the rock-salt structure—a

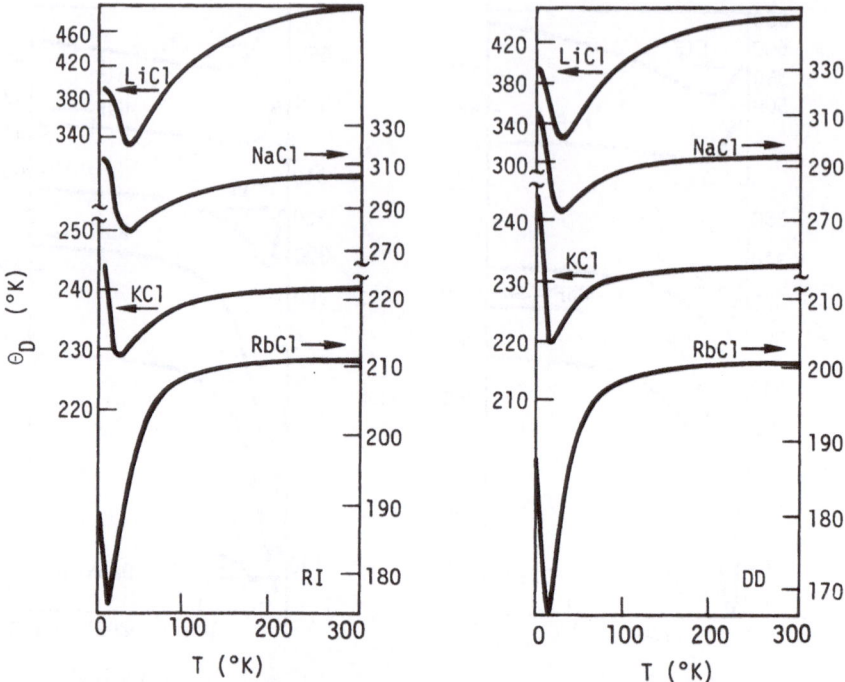

FIG. 20. Effective Debye temperature θ_D for the NaCl-structure alkali chlorides computed for the RI, DD, DD3N, and DD3N(AB) models (see caption for Fig. 19). Low-temperature input data have been used.

result that has been confirmed by experimental work.[130,132] It can also be observed that this qualitative difference is present for all the various models, which suggests that it is a structural effect.

At this stage we present in Table VII a comprehensive listing of the input parameters we have used. The quoted elastic constants, or the compressibilities calculated from them, are values measured by ultrasonic-pulse techniques. Strictly, as Liebfried and Ludwig[133] have pointed out, one should correct these to obtain the appropriate harmonic values. However, at low temperatures such corrections should be small.

The input parameters listed in Table VII are much superior to those used in our earlier studies;[74,75] in particular, the work of Lowndes and Martin[134] has provided us with reliable values for the

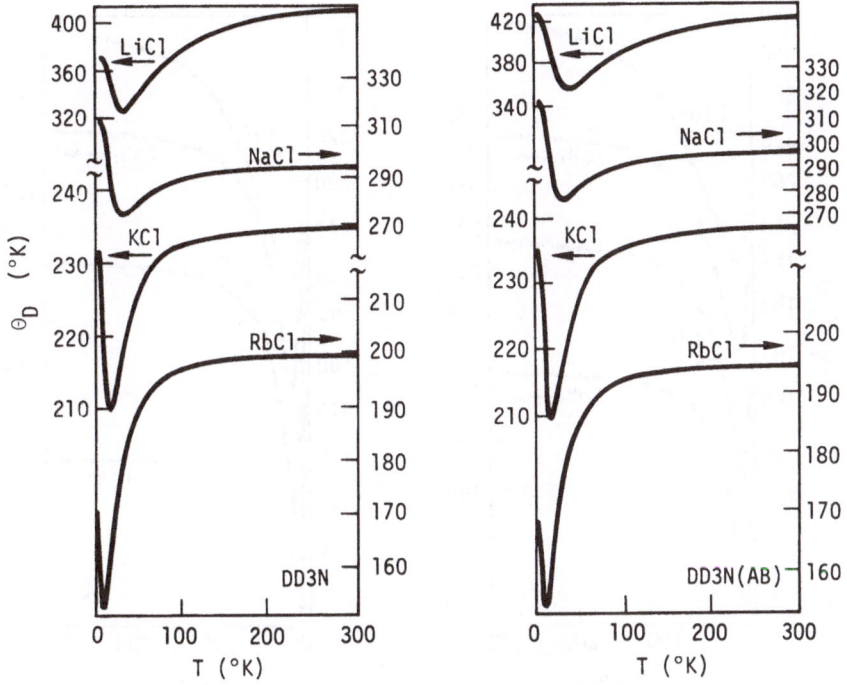

FIG. 20 (*contd.*)

static dielectric constants. The lack of such data introduced the principal uncertainty into our earlier work. The high-frequency dielectric constants were calculated directly from the Jaswal–Sharma[83] polarizabilities. If these are assumed to be approximately constant we can calculate ε_∞ at any temperature. (Data obtained by Lowndes and Martin[134] indicate this to be a reasonable assumption.) For the sake of consistency, we use these values of ε_∞ in calculating the Szigeti effective charge e^*.

The defining equations for the short-range force constants that enter into our calculations are

$$A' = \frac{12r_0^4}{\beta e^2} + \frac{4}{3}\alpha_m = \frac{v_a}{e^2}\left[\frac{d^2\Phi_{NN}(r)}{dr^2}\right]_{r=r_0} \tag{7.12a}$$

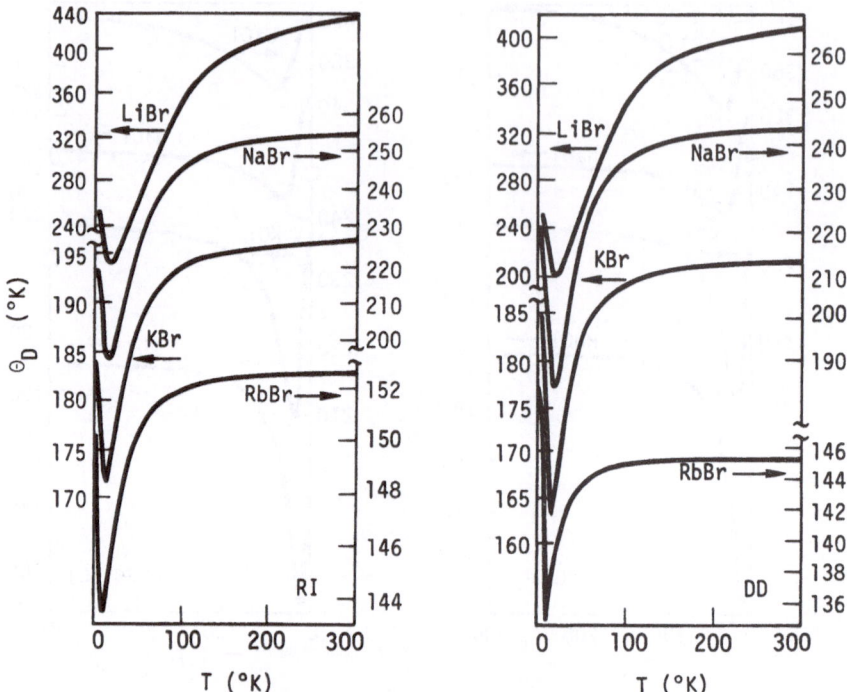

FIG. 21. Effective Debye temperature θ_D for the NaCl-structure alkali bromides computed for the RI, DD, DD3N, and DD3N(AB) models (see caption for Fig. 19). Low-temperature input data have been used.

and

$$B' = -\frac{2\alpha_m}{3} = \frac{2v_a}{e^2}\left[\frac{1}{r}\frac{d\Phi_{NN}(r)}{dr}\right]_{r=r_0} \qquad (7.12b)$$

where

$$v_a = 2r_0^3$$

Equations (7.12) are appropriate to a model for which the short-range potential Φ_{NN} is restricted to acting between first neighbors only. If we wish to allow for second-neighbor interactions,

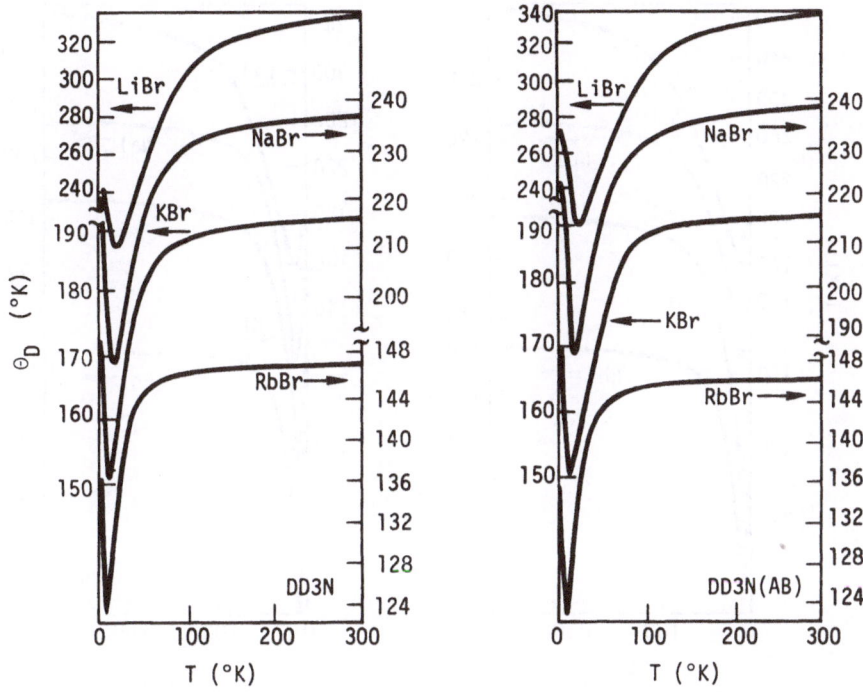

FIG. 21 (*contd.*)

the equations that determine the first-neighbor derivatives become

$$A' = 4\left[\frac{\alpha_m}{3} + \frac{r_0^4}{e^2}\left(\frac{2}{\beta^{\text{obs}}} + \frac{1}{\beta^{\text{calc}}} - \frac{8}{3}\Delta C_{44}\right)\right] \qquad (7.13\text{a})$$

and

$$B' = -4\left[\frac{\alpha_m}{6} + \frac{r_0^4}{e^2}\left(\frac{1}{\beta^{\text{obs}}} - \frac{1}{\beta^{\text{calc}}} - \frac{4}{3}\Delta C_{44}\right)\right] \qquad (7.13\text{b})$$

where

$$\Delta C_{44} = C_{44}^{\text{obs}} - C_{44}^{\text{calc}} = C_{44}^{\text{obs}} - 0.69550\,(e^2/2r_0^4)$$

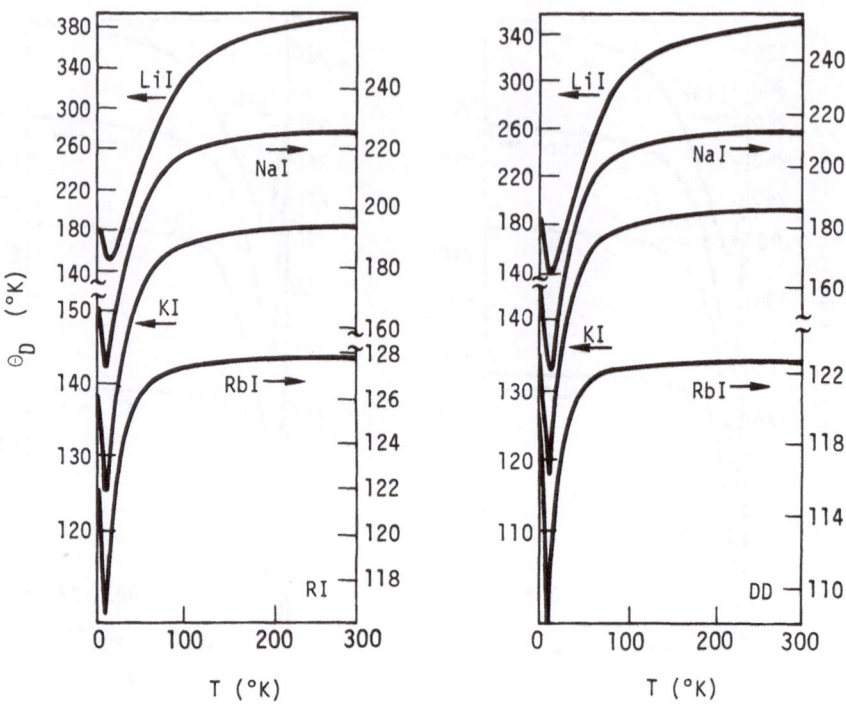

FIG. 22. Effective Debye temperature θ_D for the NaCl-structure alkali iodides computed for the RI, DD, DD3N, and DD3N(AB) models (see caption for Fig. 19). Low-temperature input data have been used.

and

$$\frac{1}{\beta^{\text{calc}}} = \frac{\bar{M}}{6r_0}\left(\frac{\varepsilon_0 + 2}{\varepsilon_\infty + 2}\right)\omega_0^2$$

$\bar{M}$ being defined as follows:

$$\bar{M} = \frac{m_1 m_2}{m_1 + m_2} = \frac{M_+ M_-}{M_+ + M_-}$$

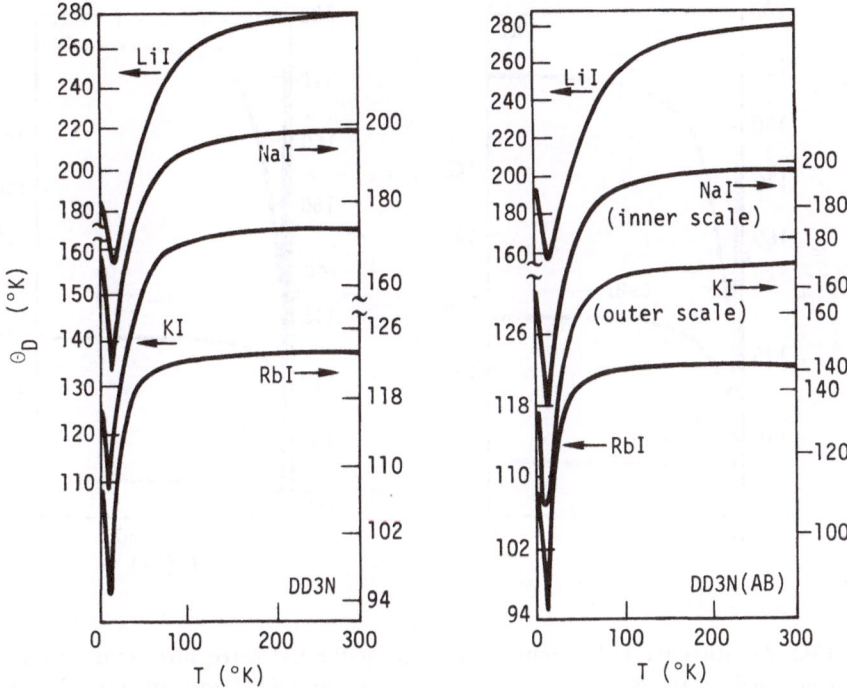

FIG. 22 (*contd.*)

(see Section 4). The equations that determine the first and second derivatives of the second-neighbor short-range interactions are

$$A'' = \frac{4r_0^4}{e^2}\left[\frac{1}{2}\left(\frac{1}{\beta^{\text{obs}}} - \frac{1}{\beta^{\text{calc}}}\right) + \frac{4}{3}\Delta C_{44}\right]$$

$$= \frac{2v_a}{e^2}\left[\frac{d^2\Phi_{NNN}(r)}{dr^2}\right]_{r=2^{1/2}r_0} \tag{7.14a}$$

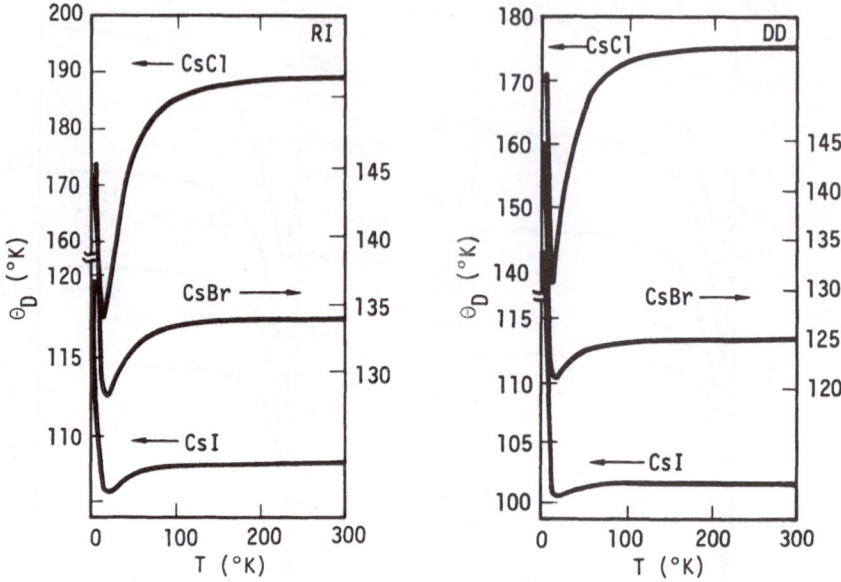

FIG. 23. Effective Debye temperature θ_D for the CsCl-structure cesium halides computed for the following models: RI (rigid ion), DD (deformation dipole with first-neighbor repulsive forces only), DD' [variation of the DD model in which the

and

$$B'' = \frac{2r_0^4}{e^2}\left[\left(\frac{1}{\beta^{obs}} - \frac{1}{\beta^{calc}}\right) - \frac{4}{3}\Delta C_{44}\right]$$

$$= \frac{2v_a}{e^2}\left[\frac{1}{r}\frac{d\Phi_{NNN}(r)}{dr}\right]_{r=2^{1/2}r_0} \tag{7.14b}$$

However, we have no means of determining the relative magnitudes of the positive–positive and the negative–negative short-range forces. Equations (7.14) define only the sum of the two. In general, we have assumed that these interactions are significant only for the larger of the two constituent ions. This type of model is adequate as

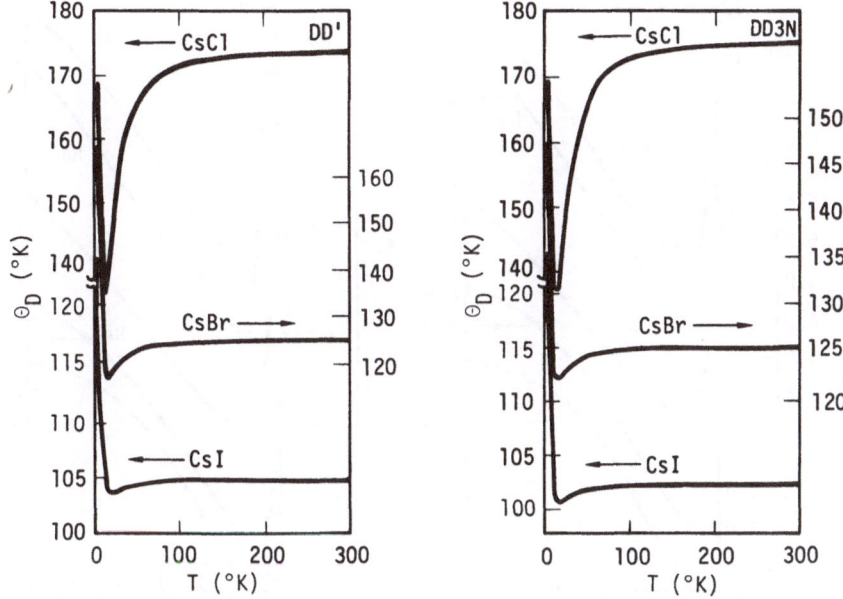

compressibility used is derived from Eq. (4.26a), i.e., the first Szigeti relation], and DD3N (deformation dipole with short-range forces extended to include closest-neighbor negative ions). Low-temperature input data have been used.

long as one assumes that the interionic forces are central. If one wishes to proceed further and allow for deviations from the Cauchy relation, $C_{12} = C_{44}$, one needs an additional parameter, such as that obtained by Woods et al.[67] by relaxing the overall equilibrium condition. This gives an additional disposable parameter

$$B_{NC} = \frac{2r_0^4}{e^2}(C_{44} - C_{12}) = \frac{2v_a}{e^2}\left[\frac{d^2\Phi_{NN}(\mathbf{r})}{dr_\perp^2}\right]_{r=r_0} \qquad (7.15)$$

where $dr_\perp$ refers to a variation perpendicular to the bond.

Alternatively, one can use the angle-bending model, first proposed by Maradudin,[92] which introduces a restoring force that resists any change in the 90° bond angles between a given ion and two

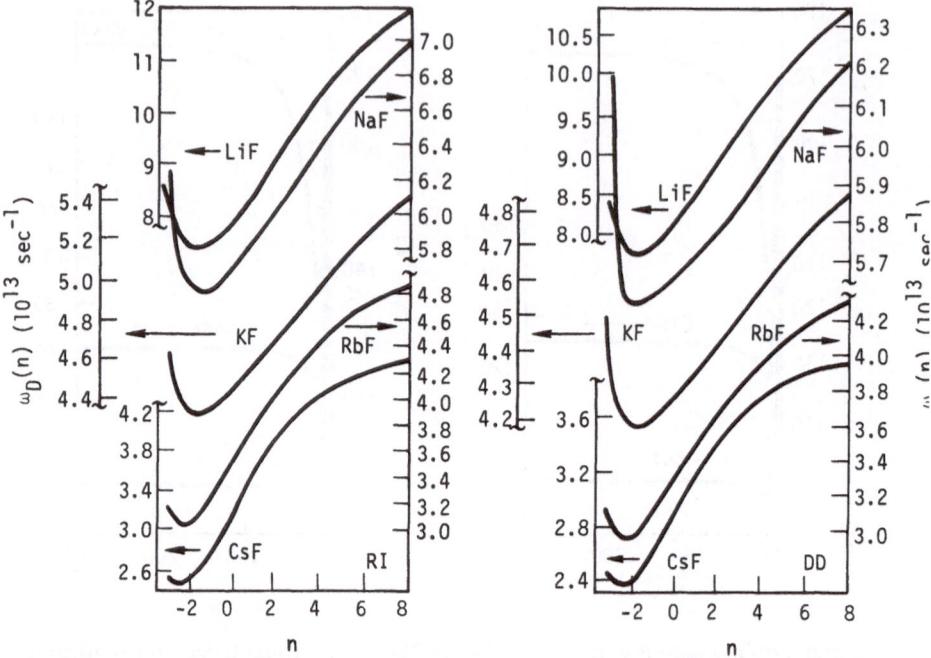

FIG. 24. Moment functions $\omega_D(n)$ [see Eq. (7.11)] for the alkali fluorides computed for the RI, DD, DD3N, and DD3N(AB) models (see caption for Fig. 19). Low-temperature input data have been used.

first neighbors. The resultant defining equations are

$$A' = 4\left\{\frac{\alpha_m}{3} + \frac{r_0^4}{e^2}\left[\left(\frac{2}{\beta^{\text{obs}}} + \frac{1}{\beta^{\text{calc}}}\right) - \frac{8}{3}\Delta C_{44} + 2(C_{44}^{\text{obs}} - C_{12}^{\text{obs}})\right]\right\}\ (7.16a)$$

$$B' = -4\left\{\frac{\alpha_m}{6} + \frac{r_0^4}{e^2}\left[\left(\frac{1}{\beta^{\text{obs}}} - \frac{1}{\beta^{\text{calc}}}\right) - \frac{4}{3}\Delta C_{44} + 2(C_{44}^{\text{obs}} - C_{12}^{\text{obs}})\right]\right\}$$

$$(7.16b)$$

$$A'' = \frac{4r_0^4}{e^2}\left[\frac{1}{2}\left(\frac{1}{\beta^{\text{obs}}} - \frac{1}{\beta^{\text{calc}}}\right) + \frac{4}{3}\Delta C_{44} - (C_{44}^{\text{obs}} - C_{12}^{\text{obs}})\right]\qquad (7.17a)$$

$$B'' = \frac{2r_0^4}{e^2}\left[\left(\frac{1}{\beta^{\text{obs}}} - \frac{1}{\beta^{\text{calc}}}\right) - \frac{4}{3}\Delta C_{44} + 2(C_{44}^{\text{obs}} - C_{12}^{\text{obs}})\right]\qquad (7.17b)$$

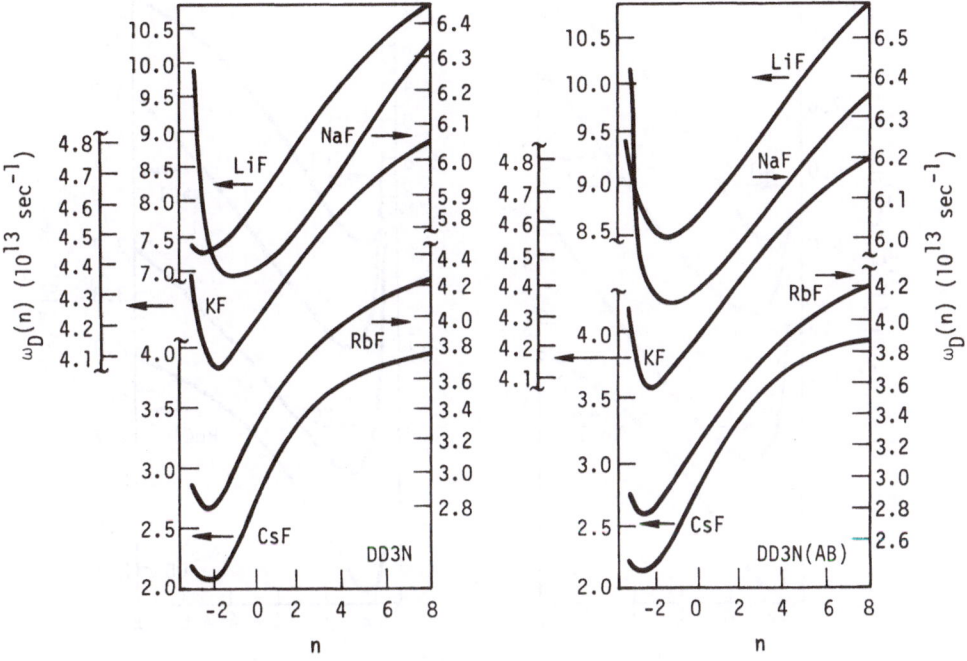

FIG. 24 (*contd.*)

The angle-bending force constant is given by

$$C = (r_0^4/e^2)(C_{44}^{obs} - C_{12}^{obs}) \qquad (7.17c)$$

and

$$BC' = B' + 4C \qquad (7.17d)$$

These results are based on the assumption that the angle-bending force constants are the same for angles about a positive ion and angles about a negative ion. Once again, this assumption is not necessary and could be relaxed. However, there is no obvious way of

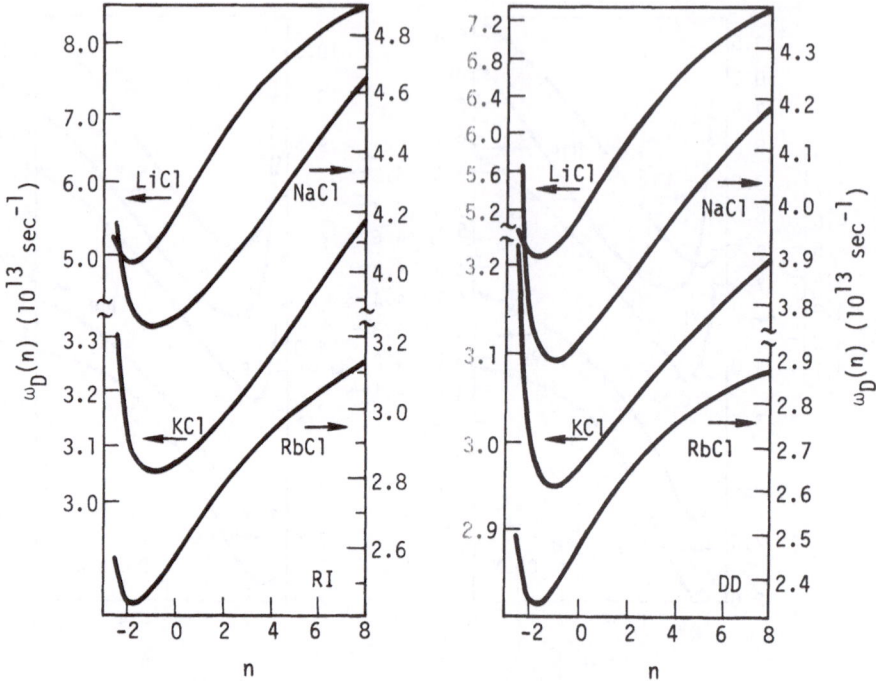

FIG. 25. Moment functions $\omega_D(n)$ [see Eq. (7.11)] for the NaCl-structure alkali chlorides computed for the RI, DD, DD3N, and DD3N(AB) models (see caption for Fig. 19). Low-temperature input data have been used.

determining the relative magnitudes of the two force constants without reference to measured dispersion curves. This model was applied by Jaswal and Hardy[135] to lithium deuteride and gives a reasonable fit to the dispersion curves measured by Verble, Warren, and Yarnell.[136]

In any given case the dispersion curves calculated with the angle-bending-force model are markedly different from those obtained by the approach of Woods *et al.*[67] Moreover, the angle-bending-force model does have the advantage that the associated potential function is rotationally invariant and the lattice is in static equilibrium.

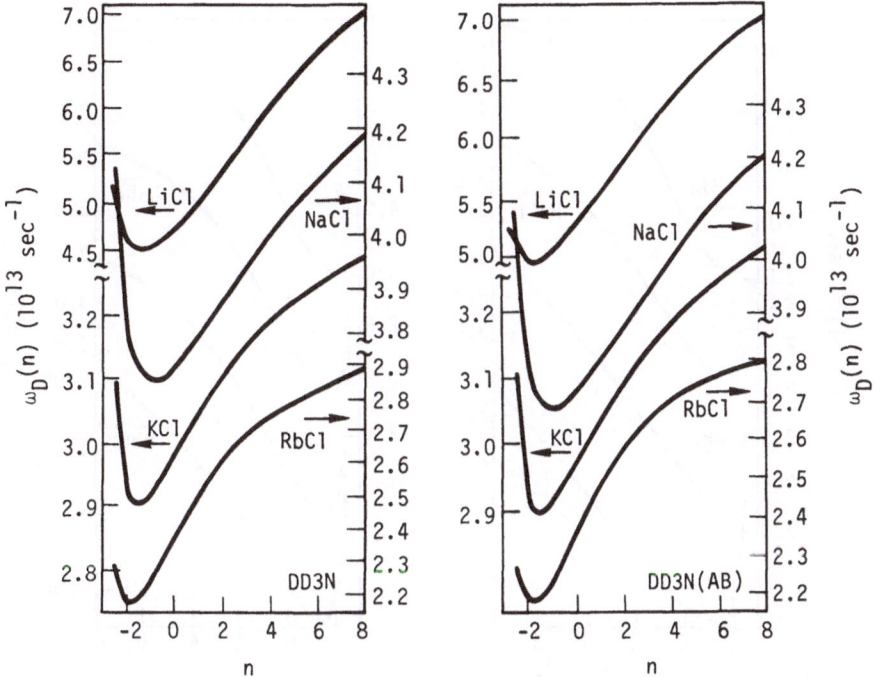

FIG. 25 (*contd.*)

The parameters involved in the dipolar part of the dynamical matrix are fixed by the second Szigeti relation [48] and the Jaswal–Sharma polarizabilities.[83]

All the survey calculations were made using a 40 × 40 × 40 mesh, which we have found to be adequate for calculating thermal data. At very low temperatures calculated values of θ_D become unreliable, as does $\omega_D(n)$ for $n < -2$. However, in both cases we know the limiting values of these functions. We can calculate the value of θ_D at absolute zero (θ_0) from the elastic constants. The moment function for $n = -3$ is $k_B\theta_0/\hbar$. Thus it is possible to interpolate both functions with good accuracy.

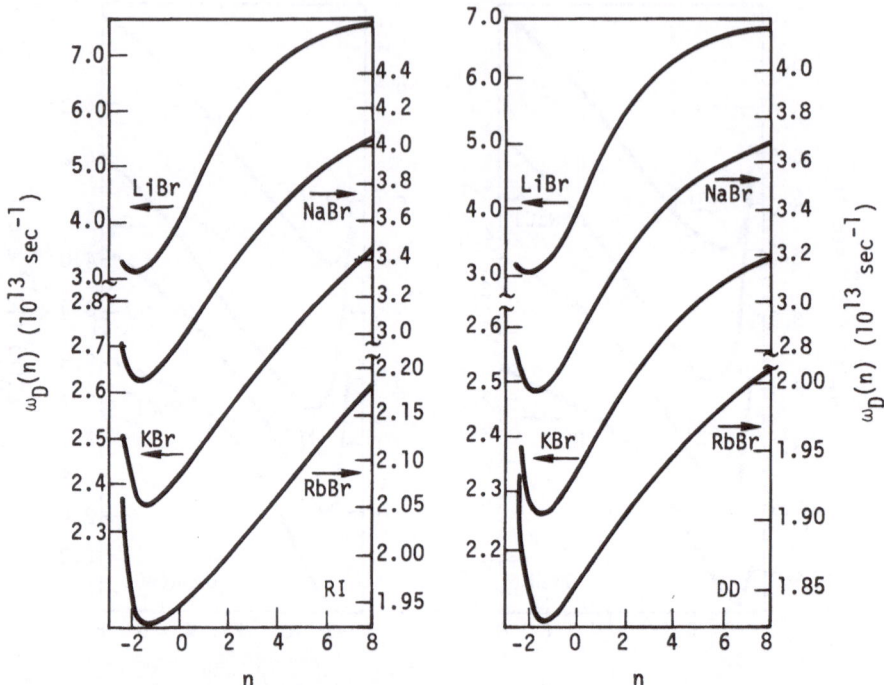

FIG. 26. Moment functions $\omega_D(n)$ [see Eq. (7.11)] for the NaCl-structure alkali bromides computed for the RI, DD, DD3N, and DD3N(AB) models (see caption of Fig. 19). Low-temperature input data have been used.

On comparing the computed thermal data with available experimental results, one finds that the deformation-dipole results are significantly better than those obtained with the rigid-ion model. The drastic changes produced by the polarization-dipole model are such as to destroy almost all agreement between theory and experiment. In some cases the use of the polarization-dipole model leads to a number of imaginary phonon frequencies. This implies that the lattice is unstable. When this situation arises, the derived thermal data have been computed by omitting all imaginary frequencies from the calculation. When the deformation dipoles are included, lattice stability is restored and, furthermore, the previously mentioned improvement in the calculated thermal data occurs. Since the

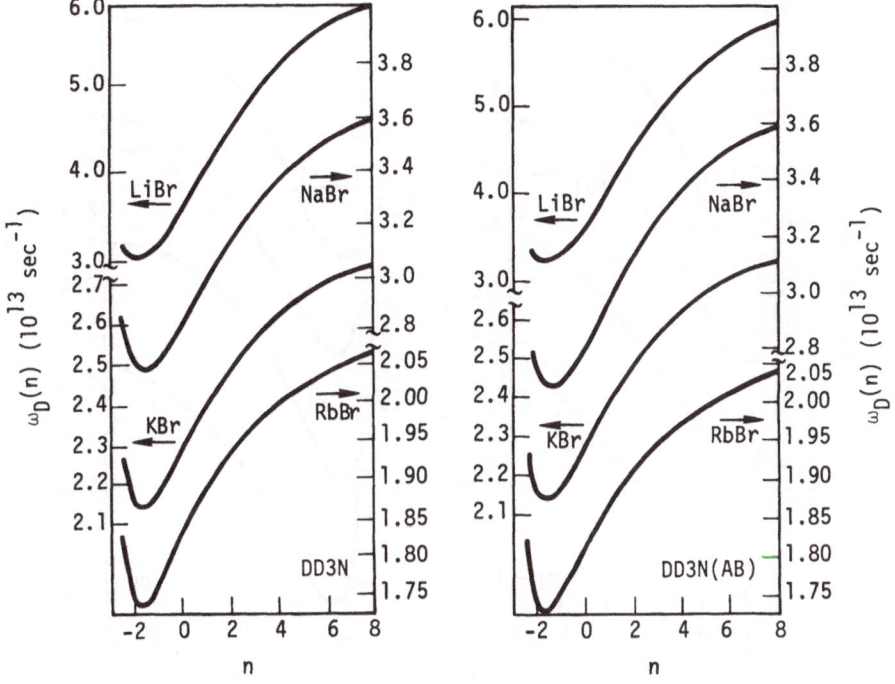

FIG. 26 (*contd.*)

polarization dipole results are so drastically different we have not displayed them explicitly in the interest of clarity.

The foregoing remarks apply to alkali halides having the rock-salt structure. For halides having the cesium chloride structure, the computed thermal data and the frequency distributions seem to be somewhat less sensitive to the dipolar model used. (For instance, in no case does one obtain lattice instability.) However, on closer inspection it becomes apparent that the nature of the dipolar model has a significant effect, particularly on the frequency spectra. For these crystals the rigid-ion model and the deformation-dipole model appear to give equally good fits to the specific-heat data. For the moment functions, the rigid-ion model appears to give

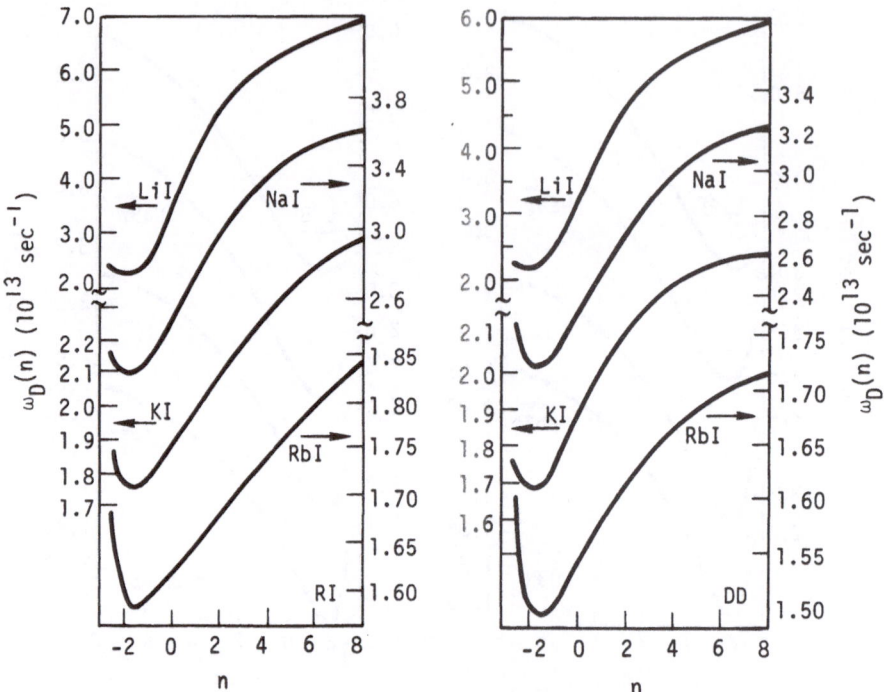

FIG. 27. Moment functions $\omega_D(n)$ [see Eq. (7.11)] for the NaCl-structure alkali iodides computed for the RI, DD, DD3N, and DD3N(AB) models (see caption of Fig. 19). Low-temperature input data have been used.

better agreement for the higher moments, whereas the deformation-dipole model appears to improve the fit for the low and negative moments.

In this section we have described the extent to which a given model can be tested by comparing computed thermal data with the corresponding experimental results. It is apparent that the tests provided by this approach are not very stringent. In the future it may be possible to refine the experimental techniques still further and extract more information. Nonetheless, it can be seen that even these limited tests are fully capable of demonstrating the absurdities of a model like the polarization-dipole model.

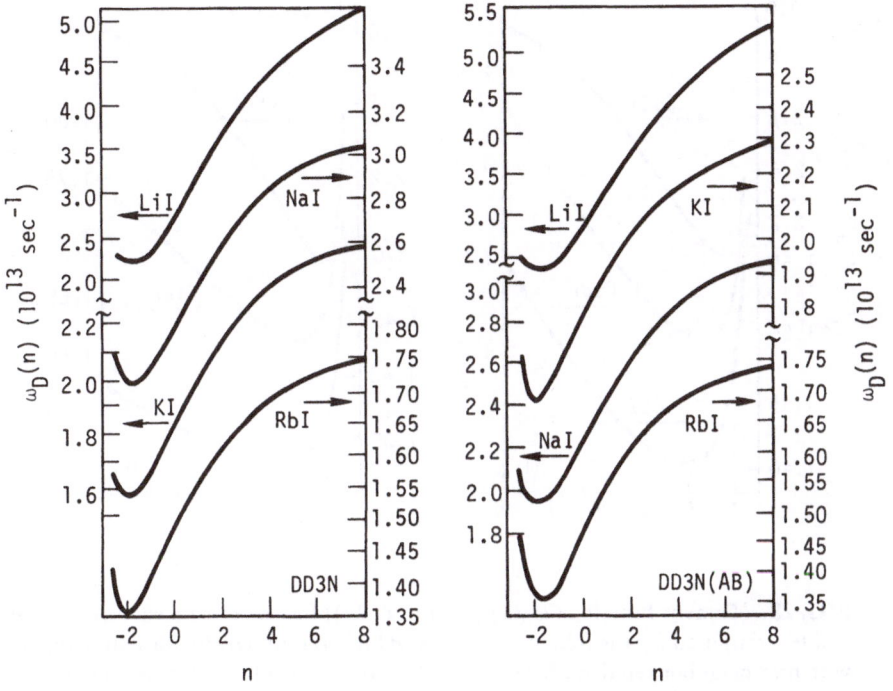

FIG. 27 (*contd.*)

8. Direct Measurement of Phonon-Dispersion Curves

This section reviews the most direct method of determining the normal modes of the crystal lattice. It is the technique that has been responsible for the revived interest in lattice dynamics over the past 20 years. The initial work on the alkali halides was performed by the Chalk River group,[67] which, under the direction of B. N. Brockhouse, developed the technique of inelastic neutron scattering into a precise experimental procedure. The basis of the technique is very simple: the use of a "probe" that will measure simultaneously the frequency and the wave vector of any lattice mode of interest. This

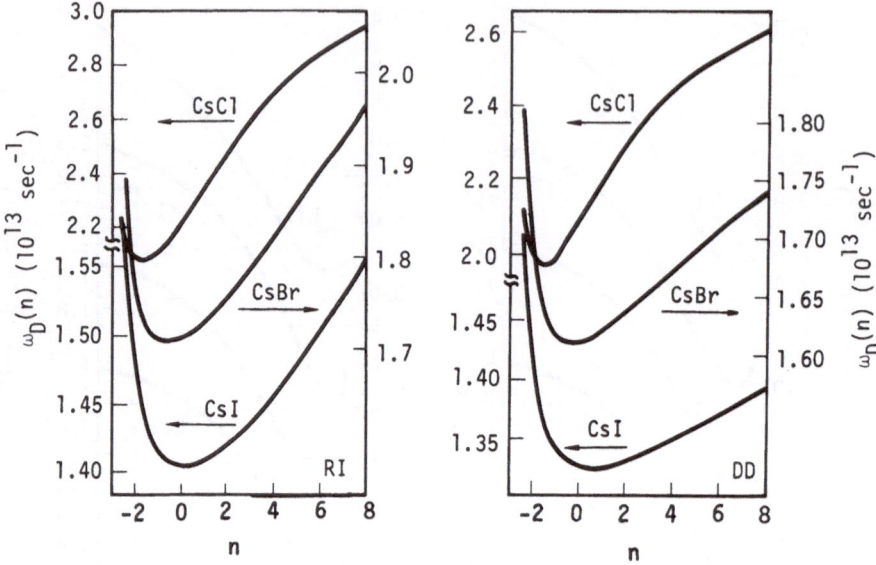

FIG. 28. Moment functions $\omega_D(n)$ [see Eq. (7.11)] for the CsCl-structure cesium halides computed for the following models: RI (rigid ion), DD (deformation dipole with first-neighbor repulsive forces only), DD′ [variation of the DD model in which

then provides a means of measuring dispersion curves directly, although it does not provide a means of measuring the whole frequency distribution directly without an enormous amount of labor. Normally one measures the dispersion curves for phonons propagating along various high-symmetry directions, usually the ⟨100⟩, ⟨110⟩, and ⟨111⟩ directions. The suitability of slow neutrons for use as such a probe results from the manner in which they are diffracted by the crystal lattice, a topic that is discussed later. Consequently, the general diffraction condition applies:

$$\mathbf{K} = \mathbf{K}_i - \mathbf{K}_s$$

$$= \mathbf{b}_h \pm \mathbf{q} \tag{8.1}$$

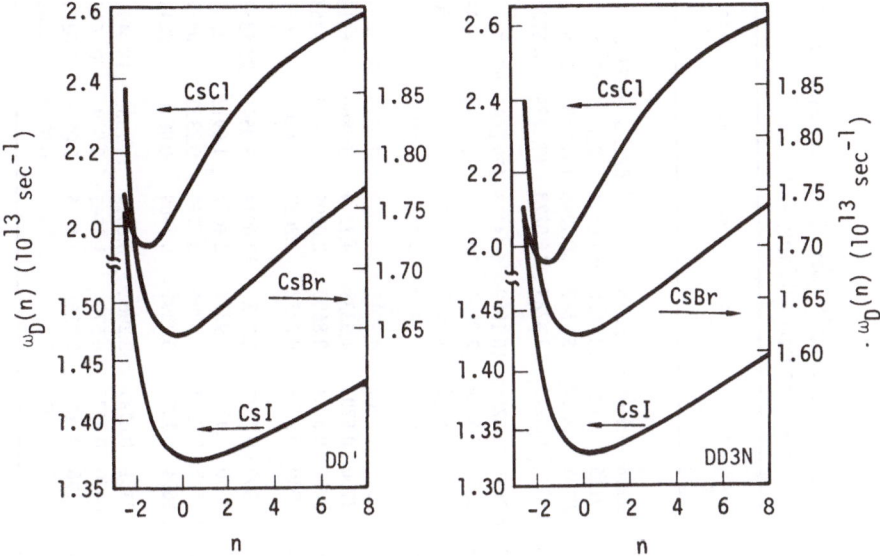

the compressibility used is derived from Eq. (4.26a), i.e., the first Szigeti relation], and DD3N (deformation dipole with short-range forces extended to include closest-neighbor negative ions). Low-temperature data have been used.

where $h\mathbf{K}$ is the momentum change of the neutron, $\mathbf{q}$ is the phonon wave vector, and $\mathbf{b}_h$ is some vector of the reciprocal lattice. This condition must be fulfilled for any inelastic-scattering process involving the participation of an elementary excitation of the solid (phonon, magnon, polariton, etc.). In this way one determines the wave vector $\mathbf{q}$ of the elementary excitation. This condition will also apply to x-ray diffraction. However, x-ray diffraction will not give a direct measurement of the energy of the elementary excitation, since this is very much smaller than the energy of the x-ray photon. Thus the change in energy in the inelastic-scattering process, which is governed by

$$\Delta E = E_i - E_s = \pm \hbar \omega \begin{pmatrix} \mathbf{q} \\ j \end{pmatrix} \tag{8.2}$$

TABLE VI. Calculated Moments of the Frequency Distributions for the Alkali Halides (10^{13n+m} $\sec^{-n}$): Deformation-Dipole Model Including Next-Nearest-Neighbor Short-Range Forces; Low-Temperature Input Data

μ_n:	$\mu_{-3.0}$	$\mu_{-2.5}$	$\mu_{-2.0}$	$\mu_{-1.5}$	$\mu_{-1.0}$	$\mu_{-0.5}$	$\mu_{0.5}$	$\mu_{1.0}$	$\mu_{2.0}$	$\mu_{3.0}$	$\mu_{4.0}$	$\mu_{6.0}$	$\mu_{8.0}$
m:	-1	-1	-1	-1	-1	-1	0	0	1	1	1	2	3
Crystal													
LiF	0.3675	0.3979	0.5745	1.0099	2.0078	4.3391	2.4339	6.2188	4.6106	39.2986	371.508	4023.75	49794.9
LiCl	1.1329	1.1382	1.4132	2.0272	3.2217	5.4894	1.9277	3.9115	1.8421	10.0658	61.6414	290.096	1620.84
LiBr	4.0472	3.2534	3.2207	3.6844	4.6900	6.5488	1.6667	3.0141	1.2032	5.7647	30.7746	104.907	409.839
LiI	9.9414	6.9982	5.9510	5.7856	6.2583	7.4989	1.4858	2.4385	0.8273	3.4289	16.0228	43.3729	138.573
NaF	0.5330	0.6162	0.8848	1.4629	2.6380	5.0338	2.0492	4.3075	2.0255	10.1999	54.4113	179.260	694.742
NaCl	1.9065	1.8102	2.1075	2.8044	4.0629	6.2378	1.6593	2.8334	0.8858	2.9803	10.6131	15.2739	24.8585
NaBr	5.5705	4.4226	4.2516	4.6609	5.6070	7.2570	1.4562	2.2252	0.5833	1.7073	5.3640	6.0069	7.4566
NaI	13.5324	9.3293	7.6845	7.1766	7.3790	8.2556	1.3040	1.8143	0.4072	1.0376	2.8402	2.4000	2.2503
KF	1.5654	1.5045	1.7830	2.4375	3.6653	5.8987	1.7674	3.2321	1.1707	4.5911	19.0891	37.6375	85.3609
KCl	4.4657	3.5731	3.5457	4.0771	5.1677	7.0098	1.4834	2.2700	0.5703	1.5333	4.3219	3.7869	3.6518
KBr	11.3070	7.7403	6.5198	6.3532	6.8635	8.0232	1.3145	1.8055	0.3764	0.8600	2.0805	1.3437	0.9363
KI	29.4651	17.2056	12.1941	9.9553	9.0781	9.1170	1.1864	1.5030	0.2778	0.5752	1.2620	0.6603	0.3671
RbF	5.2556	4.1022	3.9011	4.2731	5.2059	6.9343	1.5470	2.5399	0.7827	2.7127	10.0934	15.9545	28.5868
RbCl	11.2764	7.6140	6.3590	6.1774	6.6939	7.8997	1.3412	1.8850	0.4122	0.9838	2.4697	1.6912	1.2392
RbBr	22.1812	13.3289	9.9897	8.7443	8.5173	8.9625	1.1676	1.4127	0.2235	0.3794	0.6744	0.2333	0.0879
RbI	47.0785	24.9636	16.3824	12.5298	10.7168	0.9998	1.0578	1.1694	0.1571	0.2291	0.3509	0.0896	0.0245
CsF	12.1382	8.5352	7.0573	6.5701	6.7802	7.7818	1.4343	2.2581	0.6747	2.2962	8.2814	11.7533	18.0491
CsCl	13.9920	9.4375	7.7863	7.3306	7.5750	8.4309	1.2535	1.6470	0.3176	0.6795	1.5545	0.9191	0.6015
CsBr	20.2871	13.2117	10.5385	9.5075	9.2416	9.4396	1.0904	1.2176	0.1609	0.2261	0.3336	0.0814	0.0224
CsI	33.7853	20.4398	15.0702	12.5087	11.1382	10.3807	0.9888	0.9992	0.1077	0.1231	0.1477	0.0237	0.0043

is much too small to be measurable. However, for slow neutrons, it so happens that neutrons of wavelength comparable with the lattice spacings have energies comparable with those of typical phonons.

Thus one can directly measure the change in the energy of the neutron. At finite temperatures this can be done in two ways since there is the possibility [see Eq. (8.2)] that the neutron will either lose or gain energy. In the first process the neutron creates an elementary excitation; in the second process it destroys an elementary excitation and picks up the energy itself.

We begin with a brief discussion of the theory of neutron scattering essentially along the lines given by Lomer and Low[137] in their review. In calculating the cross section for neutron scattering by a crystal, it is usually assumed (except for magnetic scattering) that the neutrons are scattered entirely by the atomic nuclei in the crystal. For the neutron–nucleus interaction it is adequate to use the Fermi pseudopotential, which is a delta-function potential of the following form:

$$V_k(\mathbf{r}) = \frac{2\pi \hbar^2}{m} b_k \delta\left[\mathbf{r} - \mathbf{X}\binom{l}{k} \right]$$ (8.3)

Then one computes the differential scattering cross section according to first-order perturbation theory within the first Born approximation. This is certainly adequate for the neutron energies under consideration. Thus we have the following formula:

$$\frac{d^2\sigma}{d\Omega\, dE} = \frac{K_s}{2\pi \hbar K_i} \sum_{\substack{ll' \\ kk'}} b\binom{l}{k} b\binom{l'}{k'} \int_{-\infty}^{+\infty} dt \, \exp\left(2\pi i \left\{ \mathbf{K} \cdot \left[\mathbf{X}\binom{l}{k},t \right) \right.\right.$$

$$\left.\left. - \mathbf{X}\binom{l'}{k'},0 \right) \right] \right\} \right) \exp(-i\omega t)$$ (8.4)

where $\mathbf{K}_i$ and $\mathbf{K}_s$ are the wave vectors of the ingoing and the outgoing neutrons, respectively, $\mathbf{K} = \mathbf{K}_i - \mathbf{K}_s$ [momentum conservation, see

TABLE VII. Experimental Input Data

Crystal	Lattice constant r_0 (10^{-8} cm)	Elastic constants (10^{11} dynes/cm²)			Ionic polarizabilities (10^{-24} cm³)		Dielectric constant ϵ_0	Infrared dispersion frequency ω_0 (10^{13} sec⁻¹)
		C_{11}	C_{12}	C_{44}	α_+	α_-		
(a) Data at 0-4°K Appropriate to Low-Temperature Model Calculations								
LiF	2.0026	12.46	4.24	6.49	0.029	0.858	8.50	5.990
LiCl	2.5462	5.860	2.086	2.671	0.029	2.946	10.83	4.163
LiBr	2.7216	4.721	1.590	2.052	0.029	4.090	11.95	3.522
LiI	2.9610	3.49	1.319	1.50	0.029	6.114	15.28	2.854
NaF	2.3040	10.85	2.290	2.899	0.290	0.858	4.73	4.935
NaCl	2.7978	5.733	1.123	1.331	0.290	2.946	5.43	3.353
NaBr	2.9601	4.800	0.986	1.070	0.290	4.090	5.78	2.750
NaI	3.2044	3.761	0.798	0.781	0.290	6.114	6.62	2.336
KF	2.6522	7.585	1.473	1.293	1.133	0.858	5.11	3.795
KCl	3.1223	4.832	0.54	0.663	1.133	2.946	4.49	2.844
KBr	3.2711	4.180	0.560	0.525	1.133	4.090	4.52	2.317
KI	3.4998	3.38	0.22	0.368	1.133	6.114	4.66	2.062
RbF	2.7959	6.544	1.122	0.944	1.679	0.858	5.99	3.070
RbCl	3.2727	4.499	0.676	0.497	1.679	2.946	4.53	2.392
RbBr	3.4195	3.863	0.474	0.409	1.679	4.090	4.51	1.827
RbI	3.6361	3.210	0.36	0.292	1.679	6.114	4.55	1.545
CsF	2.9827	6.068	0.976	0.787	2.742	0.858	7.27	2.524
CsCl[a]	4.0563	3.560	1.194	1.080	2.742	2.946	6.68	1.978
CsBr[a]	4.2458	3.350	1.025	1.002	2.742	4.090	6.38	1.479
CsI[a]	4.5124	2.737	0.793	0.825	2.742	6.114	6.29	1.243

TABLE VII (*contd.*)

Crystal	Lattice constant r_0 (10^{-8} cm)	Elastic constants (10^{11} dynes/cm^2)			Ionic polarizabilities (10^{-24} cm^3)		Dielectric constant ε_0	Infrared dispersion frequency ω_0 (10^{13} sec^{-1})
		C_{11}	C_{12}	C_{44}	α_+	α_-		
(b) Data at 300°K Appropriate to Room–Temperature Model Calculations								
LiF	2.0132	11.12	4.20	6.28	0.029	0.876	9.032	5.764
LiCl	2.5698	4.830	1.894	2.469	0.029	3.005	11.989	3.812
LiBr	2.7498	3.920	1.890	1.885	0.029	4.168	13.295	3.219
LiI	3.0003	2.85	1.40	1.35	0.029	6.294	16.904	2.668
NaF	2.3173	9.630	2.459	2.794	0.285	0.876	5.094	4.637
NaCl	2.8203	4.870	1.311	1.266	0.285	3.005	5.9165	3.078
NaBr	2.9871	3.970	1.001	0.998	0.285	4.168	6.298	2.516
NaI	3.2372	3.007	0.912	0.733	0.285	6.294	7.304	2.180
KF	2.6735	6.480	1.600	1.252	1.149	0.876	5.514	3.574
KCl	3.1467	4.064	0.712	0.631	1.149	3.005	4.853	2.667
KBr	3.2999	3.463	0.581	0.507	1.149	4.168	4.914	2.140
KI	3.5331	2.71	0.45	0.364	1.149	6.294	5.115	1.917
RbF	2.8203	5.525	1.395	0.925	1.707	0.876	6.497	2.936
RbCl	3.2986	3.653	0.645	0.478	1.707	3.005	4.934	2.241
RbBr	3.4482	3.1630	0.4672	0.3840	1.707	4.168	4.884	1.691
RbI	3.6711	2.5730	0.3776	0.2790	1.707	6.294	4.955	1.408
CsF	3.0040	5.08	1.245	0.775	2.789	0.876	8.108	2.388
CsCl[a]	4.1003	3.6706	0.8848	0.8063	2.789	3.005	6.968	1.861
CsBr[a]	4.2957	3.0632	0.8076	0.7521	2.789	4.168	6.680	1.380
CsI[a]	4.5681	2.4524	0.6664	0.6283	2.789	6.294	6.600	1.165

[a] For these compounds r_0 denotes the cube cell side (i.e., the Cs$^+$–Cs$^+$ distance).

Eq. (8.1)], and $b\binom{l}{k}$ and $b\binom{l'}{k'}$ are the scattering lengths for the nuclei at positions $\mathbf{X}\binom{l}{k}$ and $\mathbf{X}\binom{l'}{k'}$.

Let

$$\mathbf{X}\left(\begin{matrix} l \\ k \end{matrix}, t\right) = \mathbf{r}\binom{l}{k} + \boldsymbol{\xi}\left(\begin{matrix} l \\ k \end{matrix}, t\right)$$

and

$$\mathbf{X}\left(\begin{matrix} l' \\ k' \end{matrix}, 0\right) = \mathbf{r}\binom{l'}{k'} + \boldsymbol{\xi}\left(\begin{matrix} l' \\ k' \end{matrix}, 0\right)$$

where we have introduced the label t to indicate the implicit time dependence of the various coordinates.

We now define a scattering length $\bar{b}_k$ for coherent scattering by

$$\bar{b}_k^2 = \left\langle b\binom{l}{k} \right\rangle_{\mathrm{av}}^2$$

where the average is taken over both spin and isotope for the ions of the sublattice k and is independent of the cell index l. We also define the square of the scattering length for incoherent scattering as

$$(b_k^i)^2 = \left\langle b\binom{l}{k}^2 \right\rangle_{\mathrm{av}} - \bar{b}_k^2$$

This is again independent of l and can be written as

$$(b_k^i)^2 = \langle b_k^2 \rangle_{\mathrm{av}} - \bar{b}_k^2$$

Substituting into Eq. (8.1) and remembering that [see Eqs. (2.3) and (2.11)]

$$\xi\left(\begin{matrix} l \\ k \end{matrix}, t\right) = \sum_{qj} Q\left(\begin{matrix} \mathbf{q} \\ j \end{matrix}\right)\left[\mathbf{e}\left(k \middle| \begin{matrix} \mathbf{q} \\ j \end{matrix}\right) \middle/ (Nm_k)^{1/2}\right]$$

$$\times \exp\left\{\pm i\left[2\pi\mathbf{q}\cdot\mathbf{r}\left(\begin{matrix} l \\ k \end{matrix}\right) - \omega\left(\begin{matrix} \mathbf{q} \\ j \end{matrix}\right)t\right]\right\}$$

while

$$Q\left(\begin{matrix} \mathbf{q} \\ j \end{matrix}\right)Q^*\left(\begin{matrix} \mathbf{q} \\ j \end{matrix}\right)\middle| n\rangle = \frac{\hbar}{2\omega\left(\begin{matrix} \mathbf{q} \\ j \end{matrix}\right)}(n + 1)\middle| n\rangle \tag{8.5a}$$

and

$$Q^*\left(\begin{matrix} \mathbf{q} \\ j \end{matrix}\right)Q\left(\begin{matrix} \mathbf{q} \\ j \end{matrix}\right)\middle| n\rangle = \frac{\hbar}{2\omega\left(\begin{matrix} \mathbf{q} \\ j \end{matrix}\right)}n\middle| n\rangle \tag{8.5b}$$

where $n = n\left(\begin{smallmatrix} \mathbf{q} \\ j \end{smallmatrix}\right)$ is the number of phonons present in the mode $\left(\begin{smallmatrix} \mathbf{q} \\ j \end{smallmatrix}\right)$, we obtain (after some manipulation[137]) for the one-phonon coherent-scattering cross section the following relationship:

$$\left(\frac{d^2\sigma}{d\Omega\,dE}\right)_{coh} = \left(\frac{2\pi}{V}\right)^3 N \sum_{qjh} \frac{K_s}{K_i}\delta\left[\hbar\omega\left(\begin{matrix} \mathbf{q} \\ j \end{matrix}\right) \pm (E_s - E_i)\right]\delta(\mathbf{K} \mp \mathbf{q} - \mathbf{b}_h)$$

$$\times \left\{\hbar\left[n\left(\begin{matrix} \mathbf{q} \\ j \end{matrix}\right) + \tfrac{1}{2} \pm \tfrac{1}{2}\right]\middle/ 2\omega\left(\begin{matrix} \mathbf{q} \\ j \end{matrix}\right)\right\}$$

$$\times \left| 2\pi \sum_k \bar{b}_k \exp\left[2\pi i\mathbf{K}\cdot\mathbf{r}\left(\begin{matrix} 0 \\ k \end{matrix}\right)\right]\right.$$

$$\left.\times \left[\mathbf{K}\cdot\mathbf{e}\left(k\middle|\begin{matrix} \mathbf{q} \\ j \end{matrix}\right)\middle/ m_k^{1/2}\right]\exp(-W_k)\right|^2 \tag{8.6a}$$

Here $\exp(-2W_k)$ is the Debye–Waller factor, V is the crystal volume, and the upper and lower signs correspond to phonon creation and destruction, respectively. In addition, there is one-phonon incoherent scattering, in which waves scattered by individual nuclei have no phase coherence and the cross section is given by

$$\left(\frac{d^2\sigma}{d\Omega\, dE}\right)_{\text{inc}} = \frac{(2\pi)^3}{V} \sum_{qj} \frac{K_s}{K_i} \delta\left[\hbar\omega\binom{\mathbf{q}}{j} \pm (E_s - E_i)\right]$$

$$\times \left[\hbar\left(n\binom{\mathbf{q}}{j} + \frac{1}{2} \pm \frac{1}{2}\right)\Big/ 2\omega\binom{\mathbf{q}}{j}\right]$$

$$\times \sum_k (b_k^i)^2 \left[\left|2\pi\mathbf{K} \cdot \mathbf{e}\left(k\Big|\begin{matrix}\mathbf{q}\\j\end{matrix}\right)\right|^2 \Big/ m_k\right] \exp(-2W_k)$$

$$(8.6b)$$

In general, there will be both coherent and incoherent components to the scattering cross section, and evidently there can also be multiphonon scattering of both kinds. Moreover, it is clear that there is no momentum-selection rule for incoherent scattering, and the corresponding selection rules for multiphonon coherent scattering are such that both types of scattering will cause a diffuse background that tends to obscure the sharp single-phonon peaks.

The Debye–Waller factor, $\exp(-2W_k)$, allows for the loss in intensity of the Bragg peak, whose intensity is given by the term in Eq. (8.4) with $\mathbf{X}(^l_k, t) = \mathbf{X}(^l_k, 0)$ and occurs when $\mathbf{K} = \mathbf{b}_h$. If the incoherent-scattering cross section is large, the crystal is very unsuitable for inelastic-neutron-scattering studies of the phonon-dispersion curves. However, it is possible to obtain some information about the frequency *spectrum* in the following manner: if the scattering is completely incoherent, Eq. (8.6b) shows that $(d^2\sigma/d\Omega\, dE)_{\text{inc}}$ provides a direct measure of the frequency distribution for a monatomic lattice with cubic symmetry.

If one of the elements in the material has a strong absorption cross section for neutrons, it is similarly an unfavorable case for study. These factors have impeded work on the alkali halides; in general, these materials have large incoherent-scattering cross sections and some strong absorptions. However, measurements have now been made for all the alkali halides (see Table VIII) except the lithium halides (excluding lithium fluoride). The first crystal to be studied was sodium iodide. It was these measurements, made by Woods, Cochran, and Brockhouse,[67] that first illustrated the necessity of using a more sophisticated dipolar model to describe the dynamics of ionic crystals.

TABLE VIII. Compilation of Published Data on Alkali Halide Dispersion-Curve Measurements

Crystal	Reference to experimental data
LiF	G. Dolling *et al.*, *Phys. Rev.* **168**, 970 (1968)
NaF	W. J. L. Buyers, *Phys. Rev.* **153**, 923 (1967)
NaCl	G. Raunio, L. Almquist, and R. Stedman, *Phys. Rev.* **178**, 1496 (1969)
	R. E. Schmunk and D. R. Winder, *J. Phys. Chem. Solids* **31**, 131 (1970)
NaBr	J. S. Reid, T. Smith, and ·W. J. L. Buyers, *Phys. Rev.* **B 1**, 1833 (1970)
NaI	A. D. B. Woods, W. Cochran, and B. N. Brockhouse, *Phys. Rev.* **119**, 980 (1960)
	A. D. B. Woods *et al.*, *Phys. Rev.* **131**, 1025 (1963)
KF	W. Bührer, *Phys. Status Solidi* **41**, 789 (1970)
KCl	G. Raunio, L. Almquist, and R. Stedman, *Phys. Status Solidi* **33**, 209 (1969)
	J. R. D. Copley, R. W. MacPherson, and T. Timusk, *Phys. Rev.* **182**, 965 (1969)
KBr	A. D. B. Woods *et al.*, *Phys. Rev.* **131**, 1025 (1963)
KI	G. Dolling *et al.*, *Phys. Rev.* **147**, 577 (1966)
RbF	G. Raunio and S. Rolandson, *J. Phys. C* **3**, 1013 (1970)
RbCl	G. Raunio and S. Rolandson, *J. Phys. C* **3**, 1013 (1970)
RbBr	S. Rolandson and G. Raunio, *J. Phys. C* **4**, 958 (1971)
RbI	G. Raunio and S. Rolandson, *Phys. Status Solidi* **40**, 749 (1970)
CsF	G. Raunio and S. Rolandson, *Phys. Status Solidi* **B52**, 6431 (1972)
CsCl	A. A. Z. Ahmad *et al.*, *Phys. Rev.* **B 6**, 3956 (1972)
CsBr	J. Daubert, *Phys. Letters* **32A**, 437 (1970)
	S. Rolandson and G. Raunio, *Phys. Rev.* **B 4**, 4617 (1971)
CsI	W. Bührer and W. Halg, *Phys. Status Solidi* **46**, 679 (1971)

Coherent scattering is subject to the two conservation laws given in Eqs. (8.1) and (8.2) and therefore provides a means of determining the phonon-dispersion curves. One can also see that the cross section depends on the structure factor

$$\sum_k \bar{b}_k \exp\left[2\pi i \mathbf{K} \cdot \mathbf{r}\binom{0}{k}\right] 2\pi \mathbf{K} \cdot \mathbf{e}\left(k \left| \begin{matrix} \mathbf{q} \\ j \end{matrix}\right.\right) \exp(-W_k) m_k^{-1/2} \quad (8.7)$$

which is a function of the eigenvectors of the normal modes. This is of interest because the eigenvectors contain information about the force constants that cannot be deduced from the frequencies alone.

The thermal diffuse scattering (TDS) of x rays is governed by almost the same equations as those given above, except that the scattering comes from the electrons. This approach provides a means of measuring the ratio Γ_α:

$$\Gamma_\alpha = \left[e_\alpha\left(1 \left| \begin{matrix} \mathbf{q} \\ j \end{matrix}\right.\right) \middle/ e_\alpha\left(2 \left| \begin{matrix} \mathbf{q} \\ j \end{matrix}\right.\right)\right]\left(\frac{m_2}{m_1}\right)^{1/2} \quad (8.8)$$

Determinations of this ratio are restricted to modes whose directions of polarization are determined by symmetry. However, results presented by Smith and co-workers[138–142] show that Γ can be strongly model dependent and that such attempts at "dynamical structure analysis" are very rewarding.

To determine the phonon frequencies by TDS measurements, one is forced into the difficult process of making absolute measurements of the intensities associated with the various TDS spots. This has been done, but often it is not possible to extract unambiguously the frequencies of the phonons since the cross section is a function of both the normal-mode frequencies and the eigenvectors. Indeed, in the case of x-ray TDS scattering, the scatterings from the various branches of the frequency spectrum for any given wave vector are superimposed, and in general it is not possible to separate them.

Therefore, in this case, one needs a model from which to calculate the TDS intensities. These can then be compared with the experimental values. Work of this sort has been carried out by Smith *et al.*[138-142] and by Eldridge and Lomer.[143]

Figures 29–33 show the measured and calculated dispersion curves for all the alkali halides for which dispersion curves have been measured. It should be remembered that in no case have theoretical parameters been fitted to the measured dispersion curves. The only input parameters are elastic constants and dielectric data.

When extended shell models have been used and a least-squares fit of the extra parameters has been made to the measured dispersion curves, there are significant changes that result in a better fit. However, quite often the values of the parameters that have to be used appear to be rather unphysical. Moreover, one can obtain nearly as good agreement by least-squares fitting a simple rigid-ion model, with extended short-range interactions, to the measured dispersion curves.[144] In fact, such a model may contain fewer disposable parameters.

In the case of lithium fluoride,[145] it is difficult to get a good fit to the measured dispersion curves without unrealistic modification of the input parameters. This supports our original contention in 1963[75] that there is, in fact, no "typical alkali halide"; each one has to be studied independently. In the case of sodium fluoride, it is interesting to see that the deformation-dipole model with next-nearest-neighbor interactions provides a significantly better fit to the measured dispersion curves than does the simple shell model. On the other hand, for the case of magnesium oxide,[79,80] the breathing-shell model appears to be best able to reproduce the measured dispersion curves. Overall, there is little to choose between the deformation-dipole model and the simple shell model as regards fit to the measured dispersion curves.

Benedek and Maradudin[101] have modified the deformation-dipole model for potassium iodide by fitting certain of the parameters

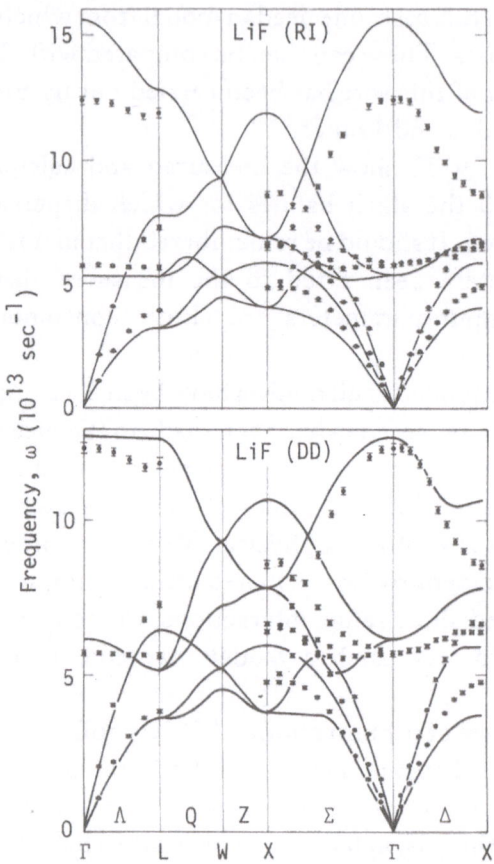

FIG. 29a. Low-temperature dispersion curves for LiF computed for models: RI (rigid ion); PD (polarization dipole: ionic polarization included, short-range deformation excluded); DD (deformation dipole with first-neighbor repulsive forces

in the model to zone-boundary frequencies. They did this to ensure that the gap in the frequency spectrum is correctly placed. Specifically, two zone-boundary frequencies were fitted in order to define the upper and lower edges of this gap. The resultant eigenfrequencies and eigenvectors were used in a calculation of the infrared absorption due to a substitutional defect (Cl^-) in potassium

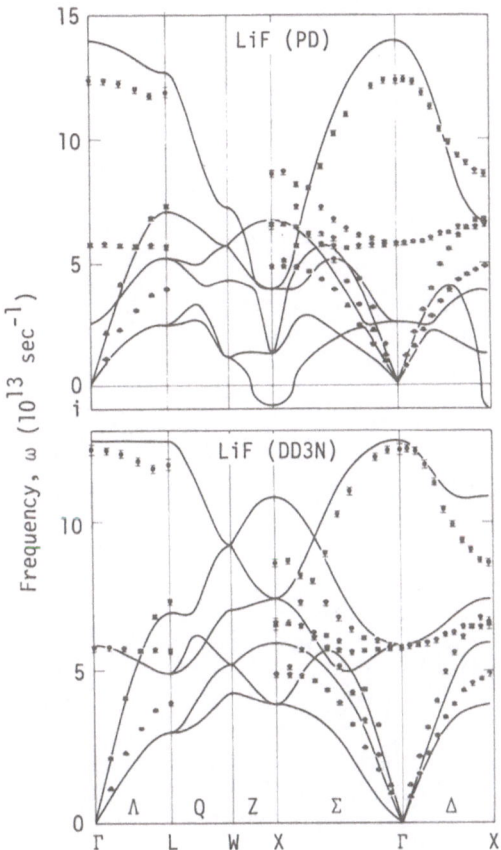

only); and DD3N (deformation dipole with short-range forces extended to include closest-neighbor negative ions). Points are 298°K experimental data (see Table VIII).

iodide, and the computed spectrum shown in Fig. 34 is in significantly better agreement with the experiment than that computed using the eigenvector and eigenfrequency data derived from the extended shell model. However, the values of the polarizabilities that were used in order to fit these zone-boundary frequencies are somewhat unphysical.

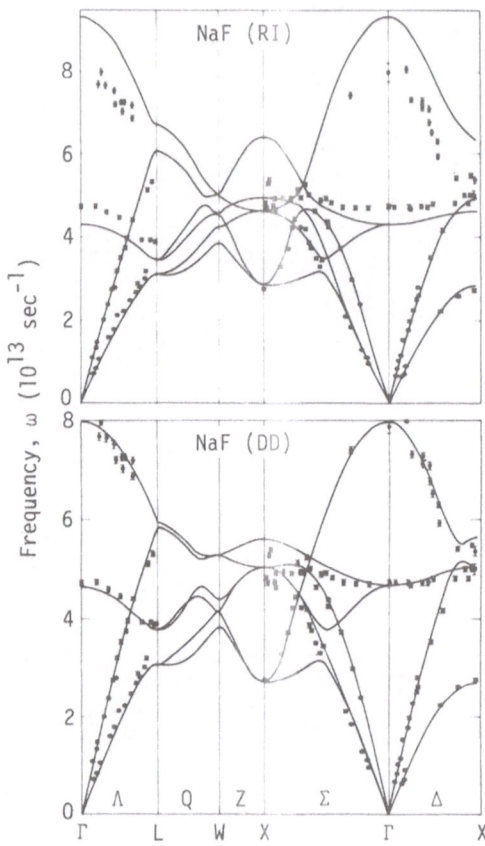

FIG. 29b. Low-temperature dispersion curves for NaF computed for the RI, PD, DD, and DD3N models (see caption for Fig. 29a). Points are 295°K experimental data (see Table VIII).

Inelastic neutron scattering has provided the first clear evidence of phononlike elementary excitations in crystal lattices. This technique permits direct studies of these excitations, and it has now become of critical importance to be able to calculate these dispersion curves. Before this technique was developed, there was a strong tendency to use highly oversimplified models. The first measurements on sodium iodide were as important as the first measurements on germanium[146] in that they revealed quite clearly that the details

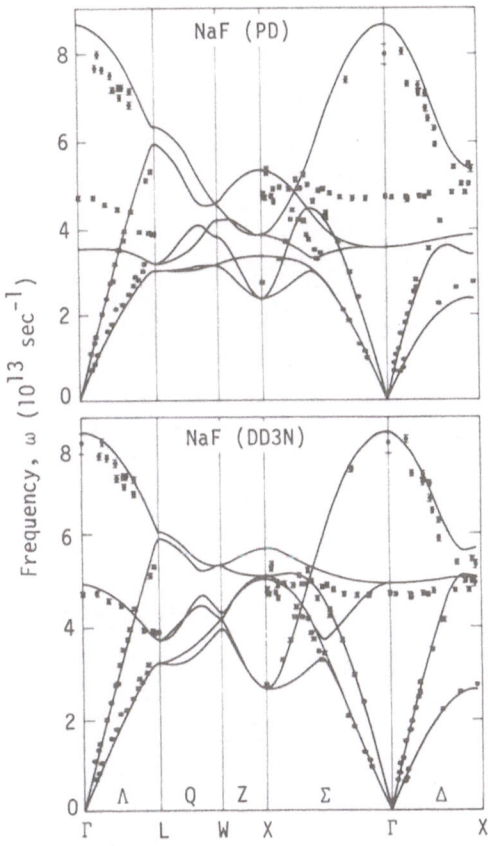

FIG. 29b (*contd.*)

of the effective potential function employed for the lattice dynamics have to be taken into account before a meaningful comparison between theory and experiment can be made. The various types of dipole models represent the best type of theoretical insight that is available at present. *A priori* calculations have yet to be carried to the point of producing the type of agreement between theory and experiment that the various models provide.[147]

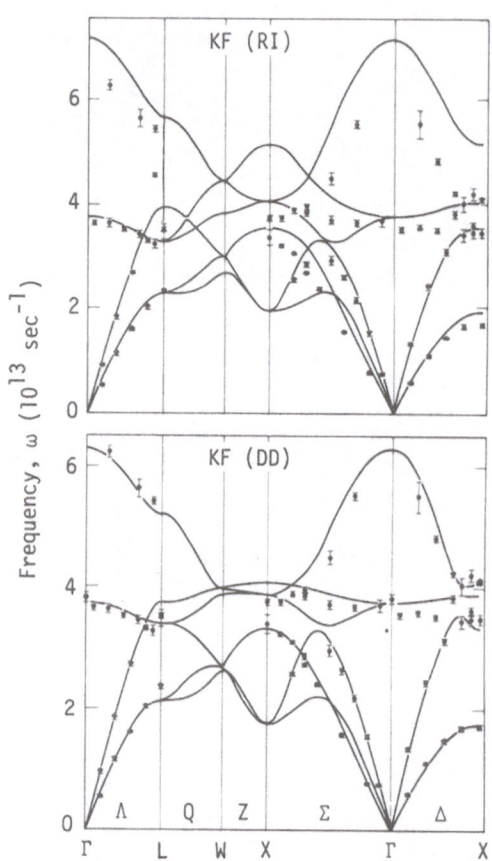

FIG. 29c. Low-temperature dispersion curves for KF computed for the RI, PD, DD, and DD3N models (see caption for Fig. 29a). Points are room-temperature experimental data (see Table VIII).

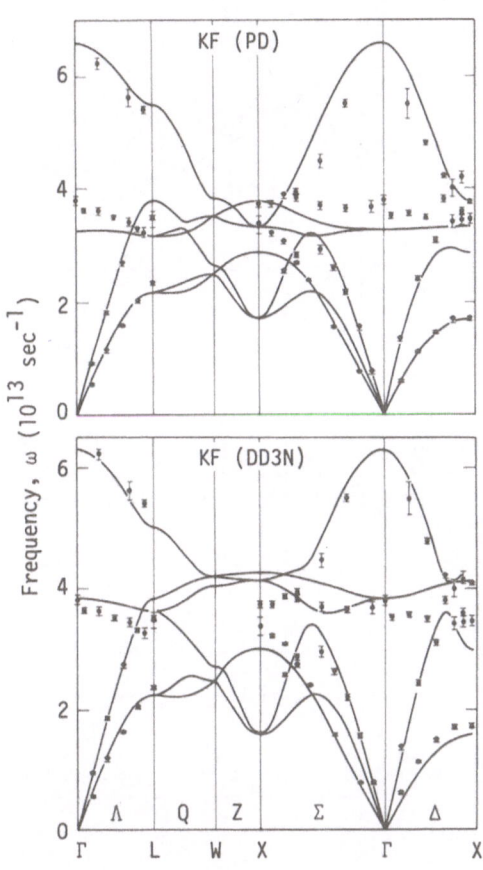

FIG. 29c (*contd.*)

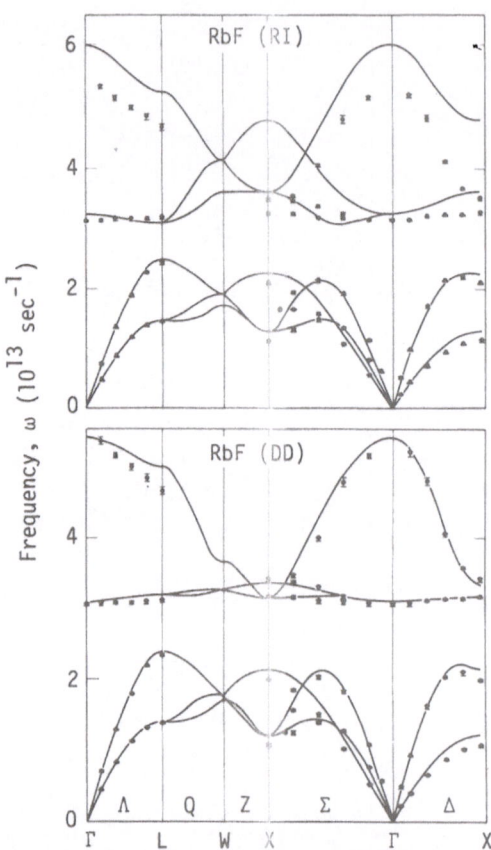

FIG. 29d. Low-temperature dispersion curves for RbF computed for the RI, PD, DD, and DD3N models (see caption for Fig. 29a). Points are 80°K experimental data (see Table VIII).

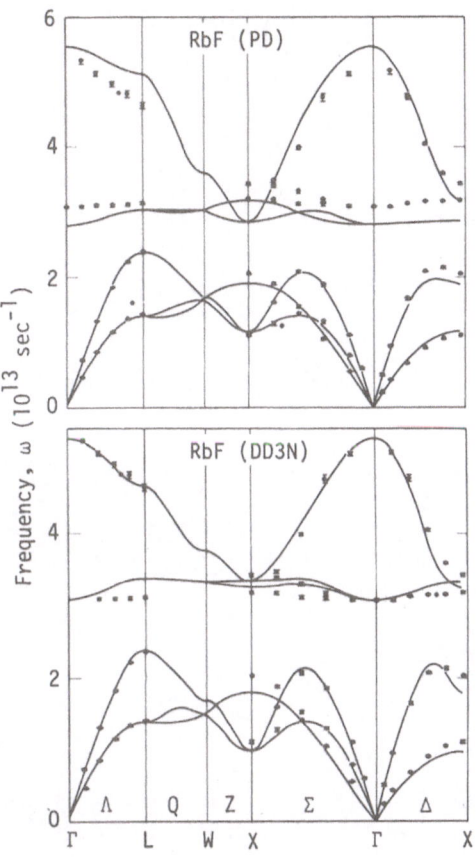

FIG. 29d (*contd.*)

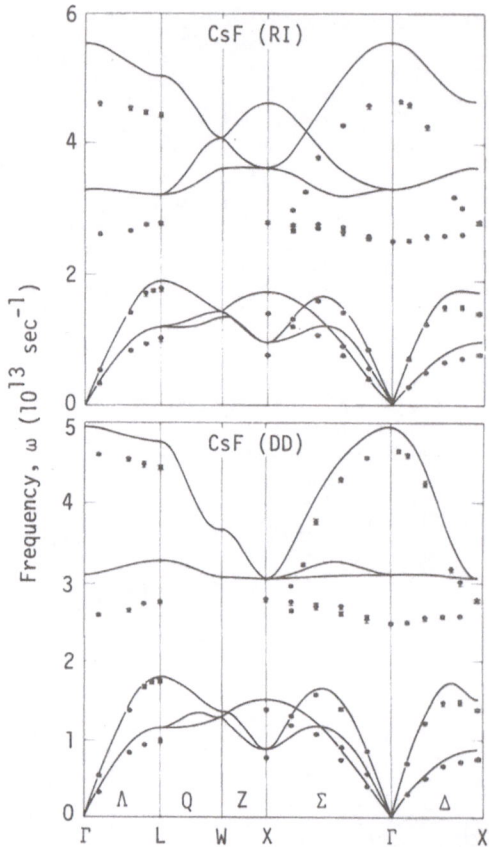

FIG. 29e. Low-temperature dispersion curves for CsF computed for the RI, PD, DD, and DD3N models (see caption for Fig. 29a). Points are 80°K experimental data (see Table VIII).

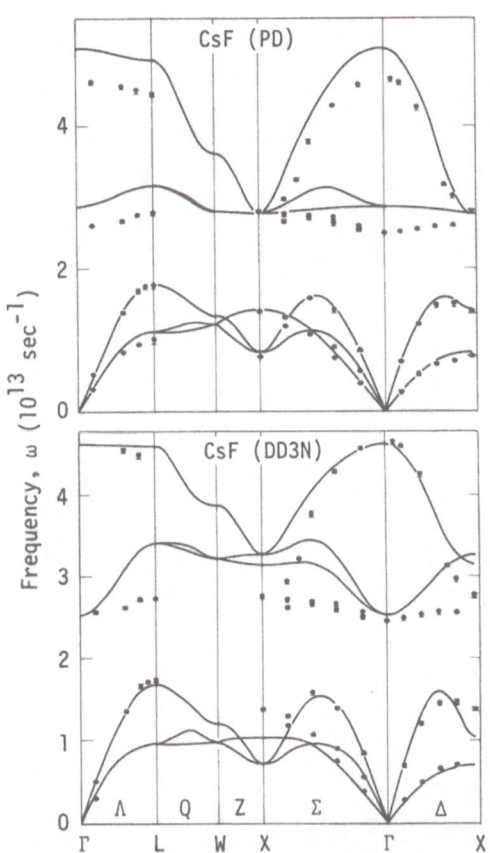

FIG. 29e (*contd.*)

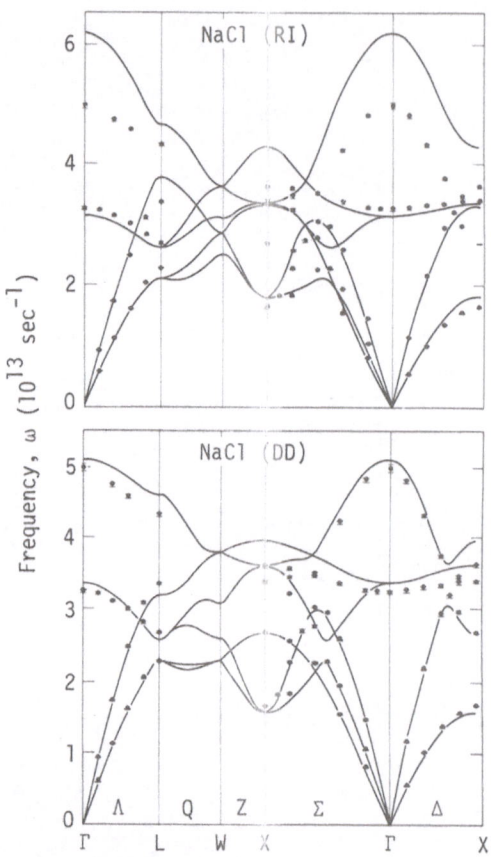

FIG. 30a. Low-temperature dispersion curves for NaCl computed for the RI, PD, DD, and DD3N models (see caption for Fig. 29a). Points are 80°K experimental data (see Table VIII).

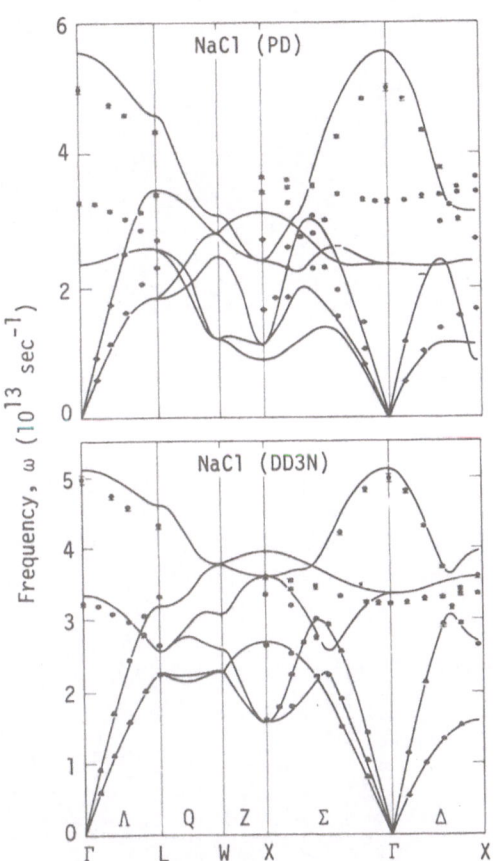

FIG. 30a (*contd.*)

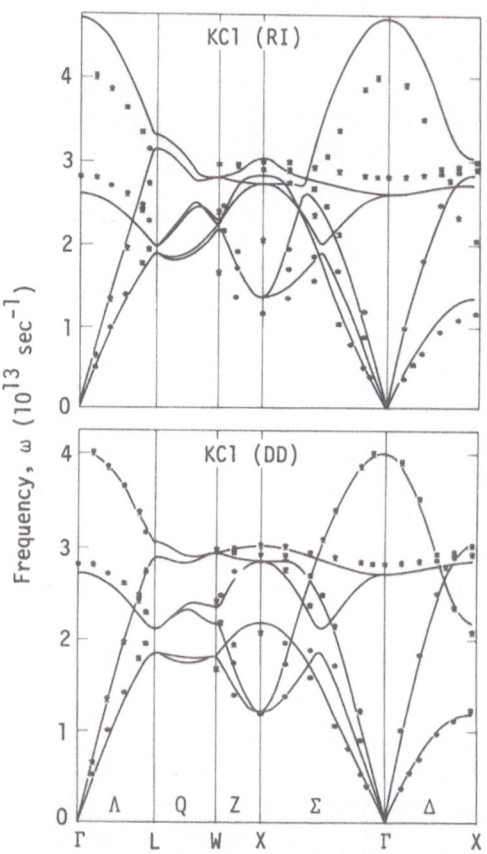

FIG. 30b. Low-temperature dispersion curves for KCl computed for the RI, PD, DD, and DD3N models (see caption for Fig. 29a). Points are 80°K experimental data (see Table VIII).

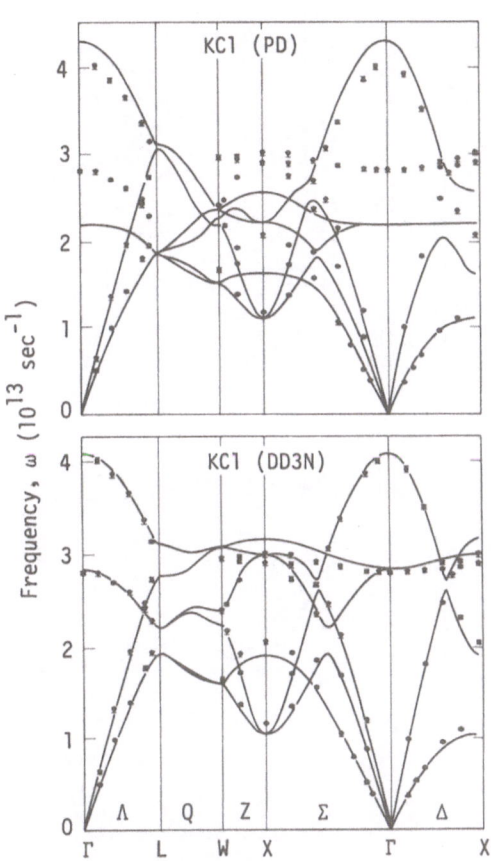

FIG. 30b (*contd.*)

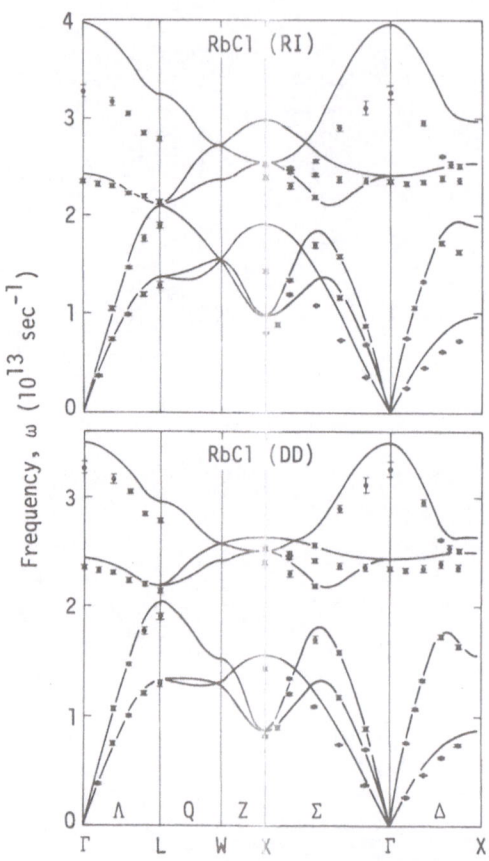

FIG. 30c. Low-temperature dispersion curves for RbCl computed for the RI, PD, DD, and DD3N models (see caption for Fig. 29a). Points are 80°K experimental data (see Table VIII).

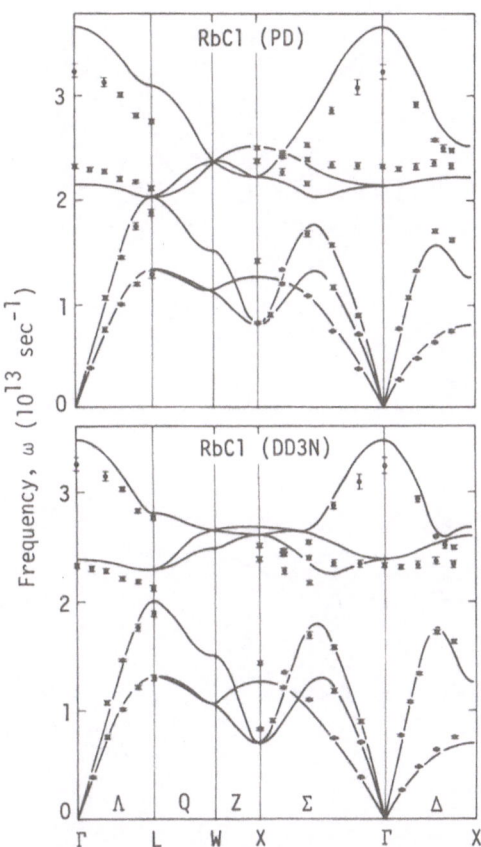

FIG. 30c (*contd.*)

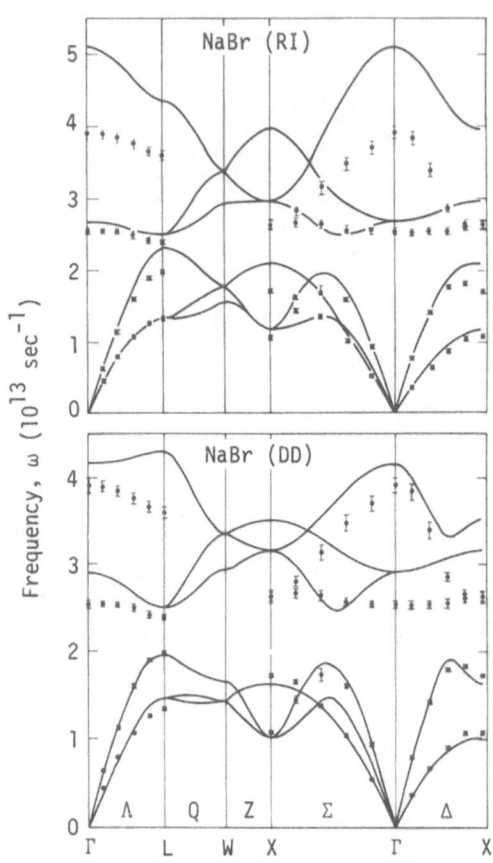

Fig. 31a. Low-temperature dispersion curves for NaBr computed for the RI, PD, DD, and DD3N models (see caption for Fig. 29a). Points are 80°K experimental data (see Table VIII).

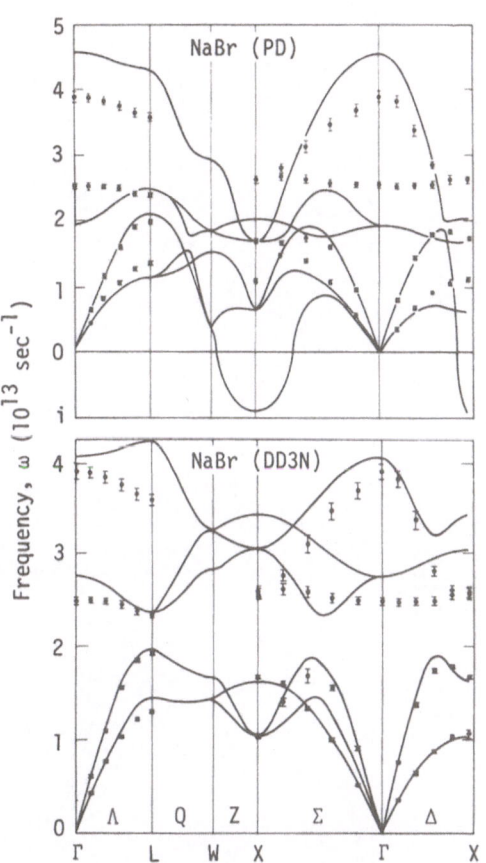

FIG. 31a (*contd.*)

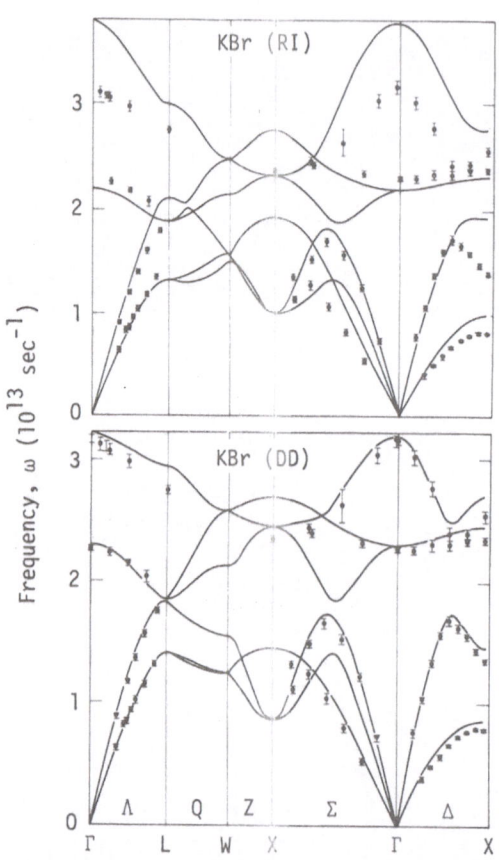

FIG. 31b. Low-temperature dispersion curves for KBr computed for the RI, PD, DD, and DD3N models (see caption for Fig. 29a). Points are 90°K experimental data (see Table VIII).

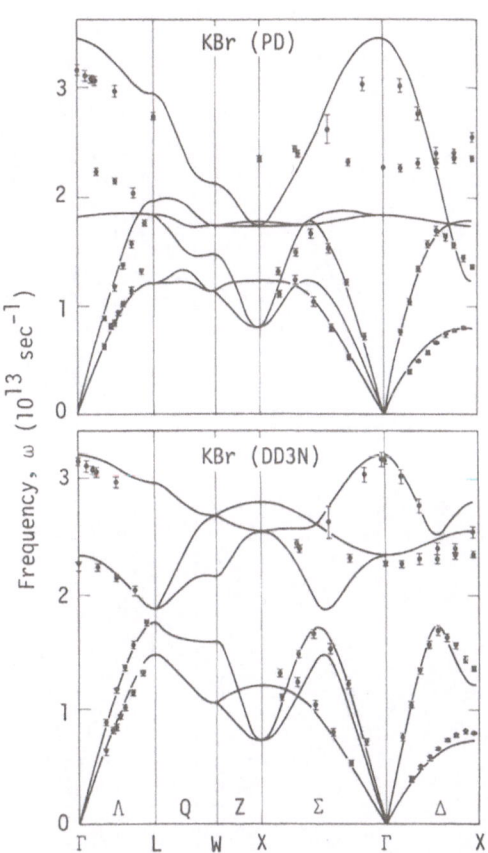

FIG. 31b (*contd.*)

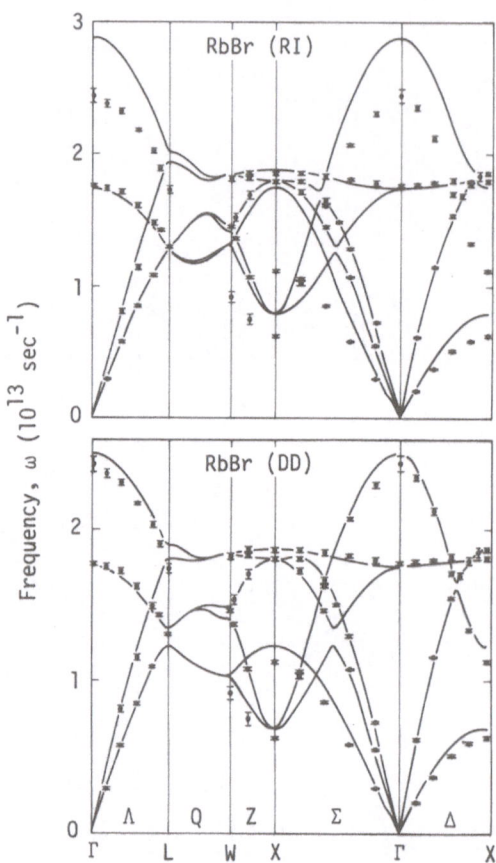

FIG. 31c. Low-temperature dispersion curves for RbBr computed for the RI, PD, DD, and DD3N models (see caption for Fig. 29a). Points are 80°K experimental data (see Table VIII).

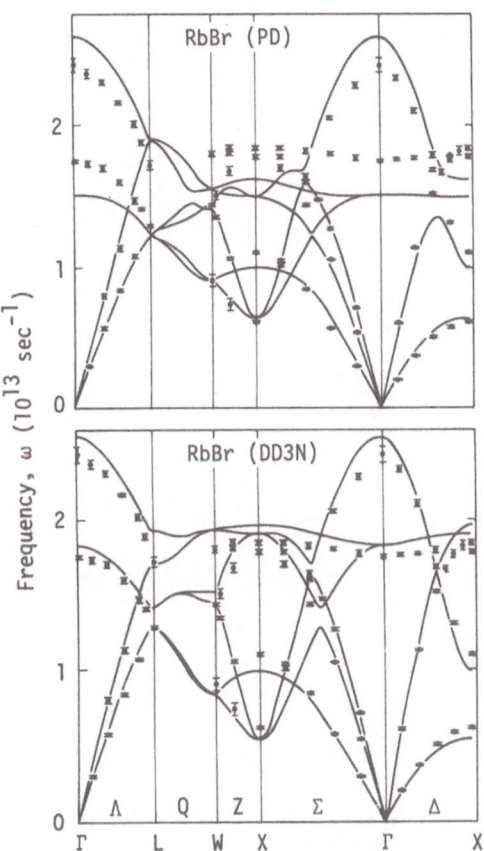

FIG. 31c (*contd.*)

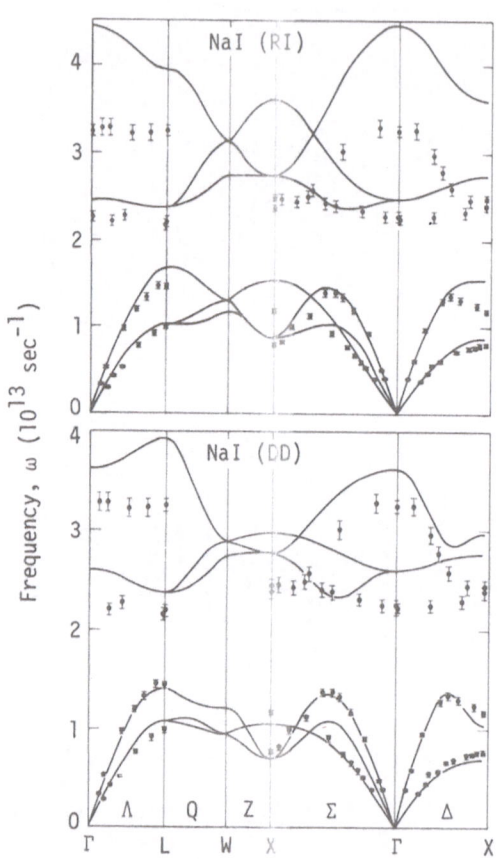

FIG. 32a. Low-temperature dispersion curves for NaI computed for the RI, PD, DD, and DD3N models (see caption for Fig. 29a). Points are 100°K experimental data (see Table VIII).

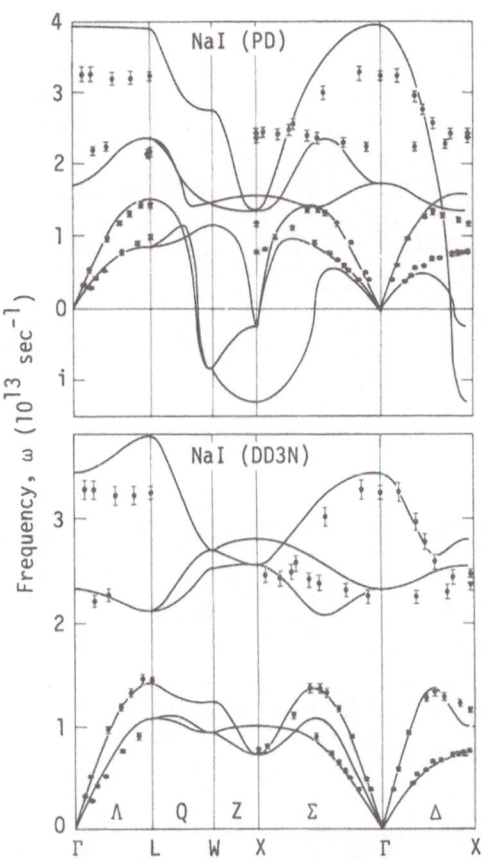

FIG. 32a (*contd.*)

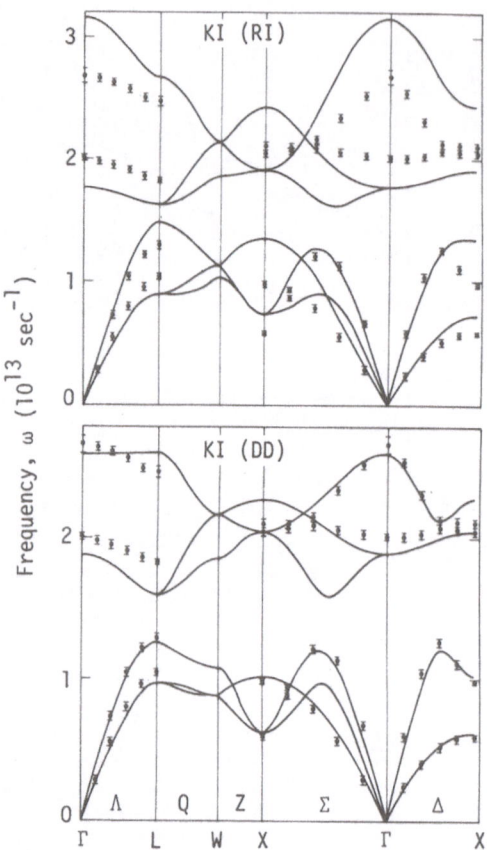

FIG. 32b. Low-temperature dispersion curves for KI computed for the RI, PD, DD, and DD3N models (see caption for Fig. 29a). Points are 90°K experimental data (see Table VIII).

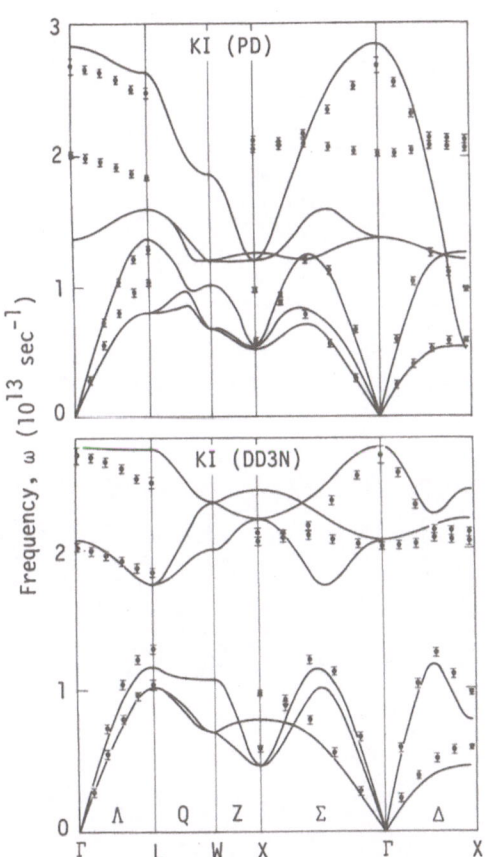

FIG. 32b (*contd.*)

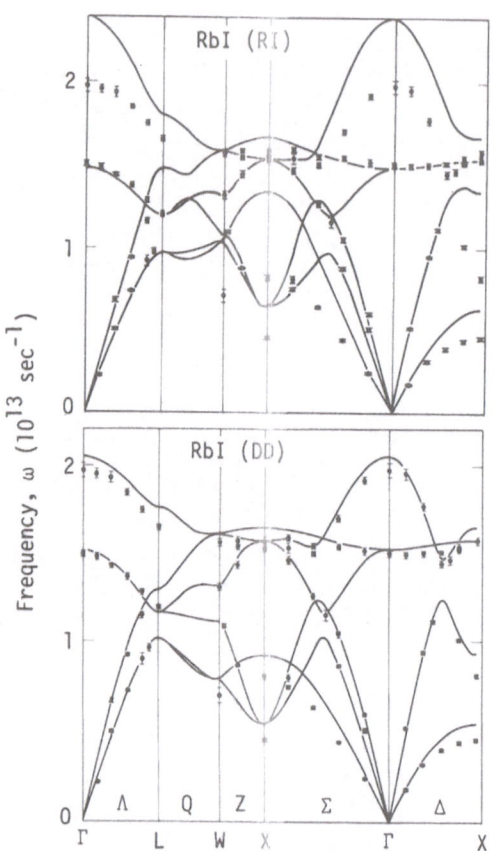

FIG. 32c. Low-temperature dispersion curves for RbI computed for the RI, PD, DD, and DD3N models (see caption for Fig. 29a). Points are 80°K experimental data (see Table VIII).

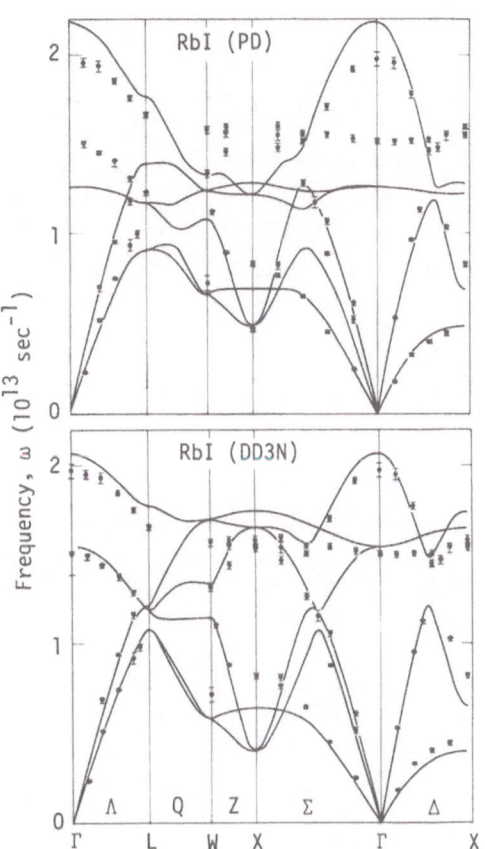

FIG. 32c (*contd.*)

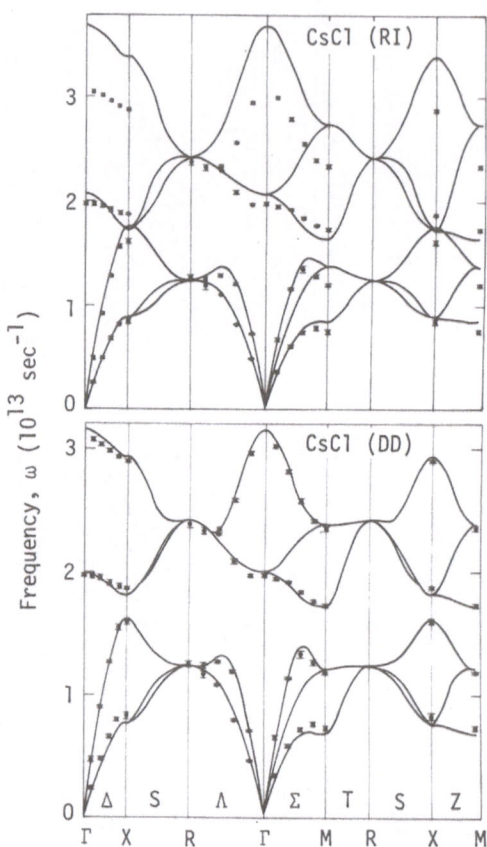

FIG. 33a. Low-temperature dispersion curves for CsCl computed for the RI, PD, DD, and DD3N models (see caption for Fig. 29a). Points are 78°K experimental data (see Table VIII).

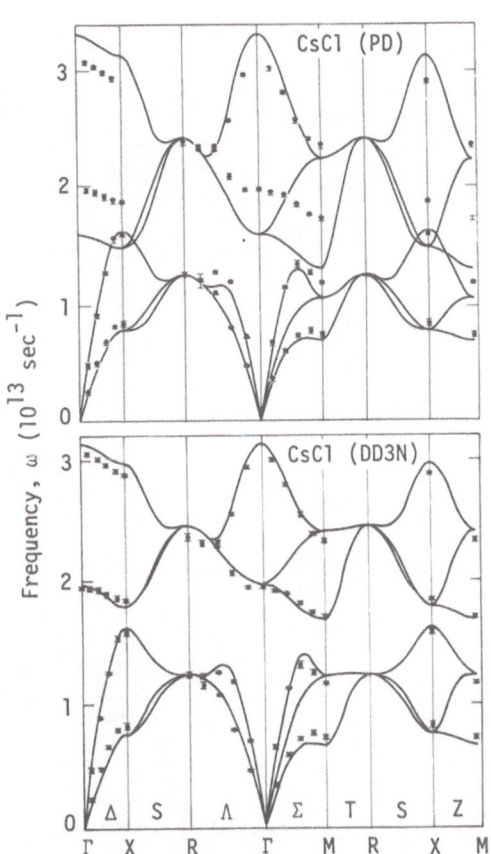

FIG. 33a (*contd.*)

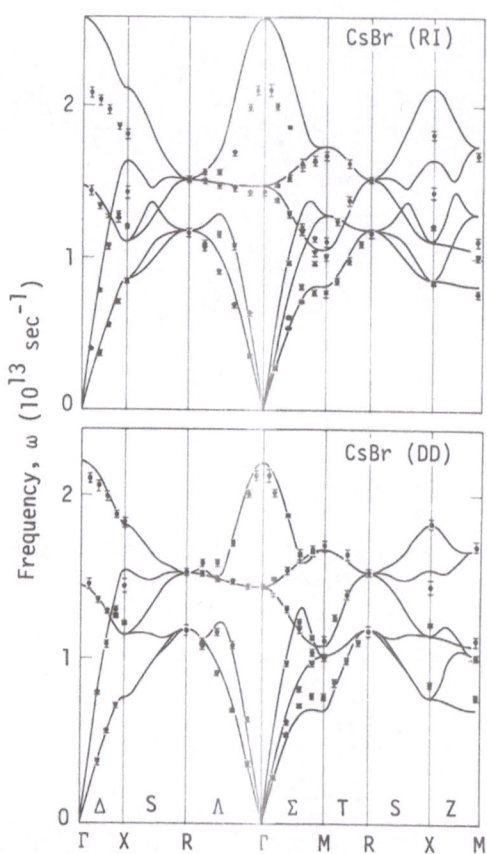

FIG. 33b. Low-temperature dispersion curves for CsBr computed for the RI, PD, DD, and DD3N models (see caption for Fig. 29a). Points are 80°K experimental data (see Table VIII).

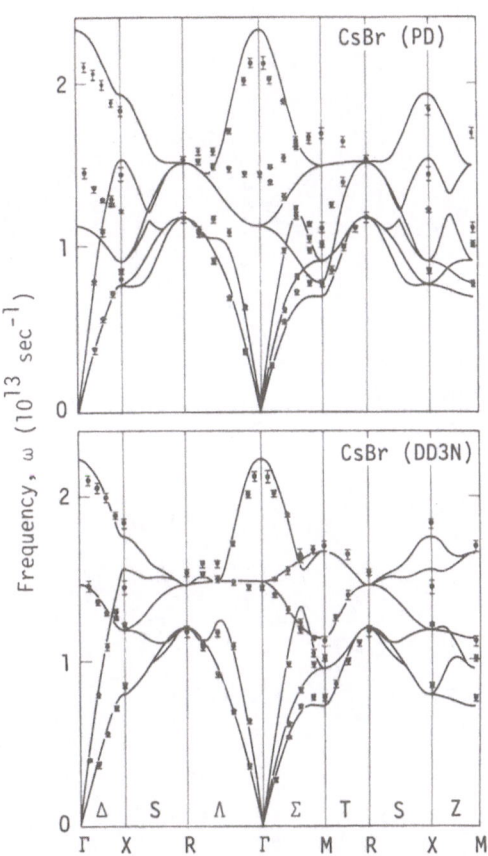

FIG. 33b (*contd.*)

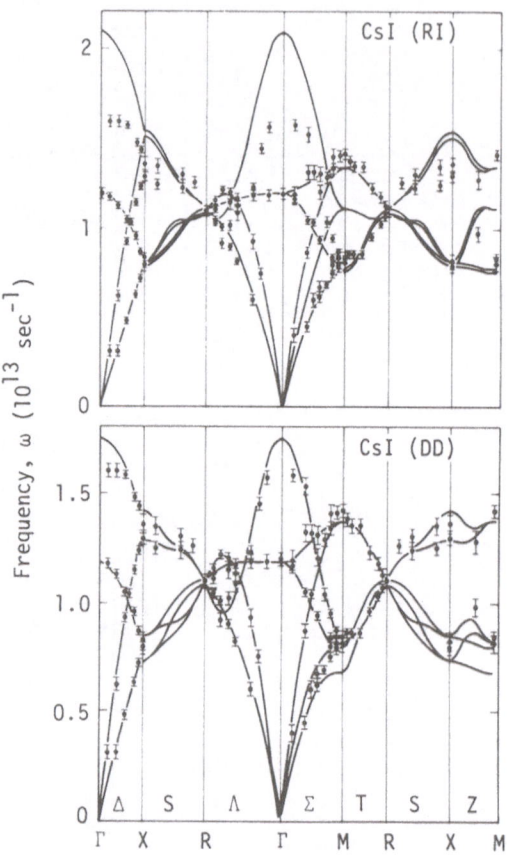

FIG. 33c. Low-temperature dispersion curves for CsI computed for the RI, PD, DD, and DD3N models (see caption for Fig. 29a). Points are room-temperature experimental data (see Table VIII).

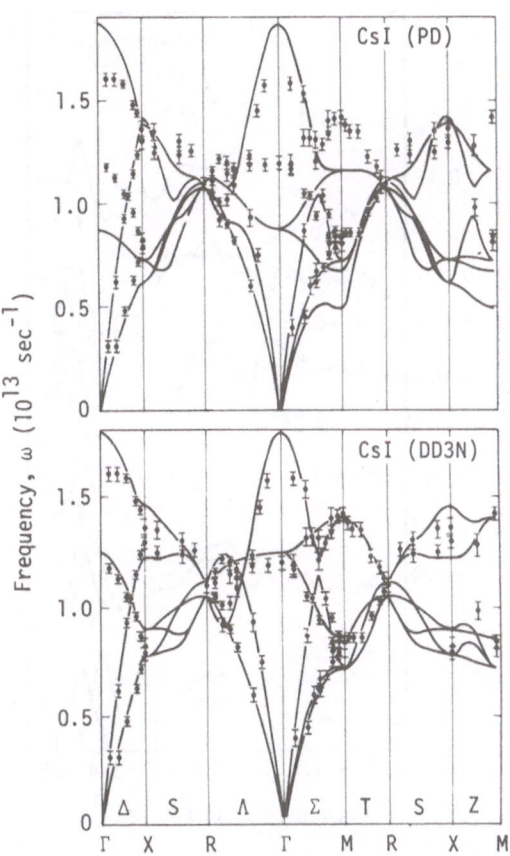

FIG. 33c (*contd.*)

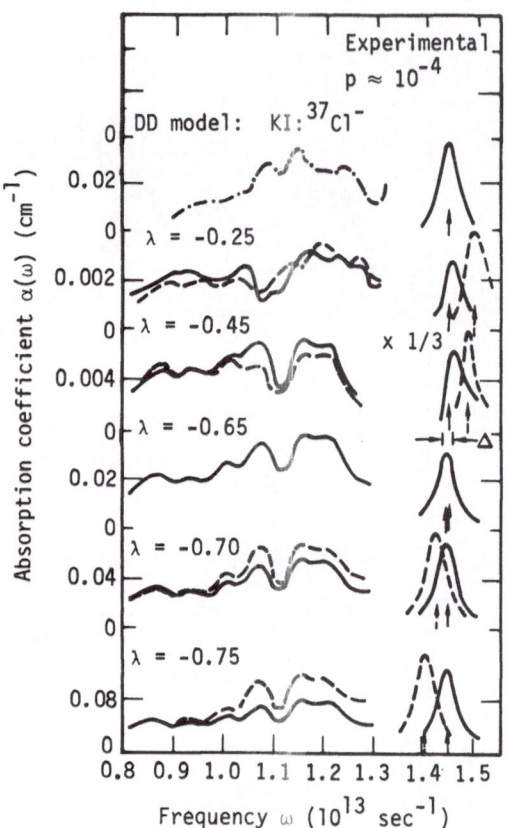

FIG. 34. Calculated impurity-activated absorption coefficients for KI containing a fraction $p = 10^{-4}$ of $^{37}Cl^-$ substituted for I^-, after G. B. Benedek and A. A. Maradudin, *J. Phys. Chem. Solids* **29**, 423 (1968). The defect is characterized by changes in the nearest-neighbor central and noncentral forces λ and λ'. The full

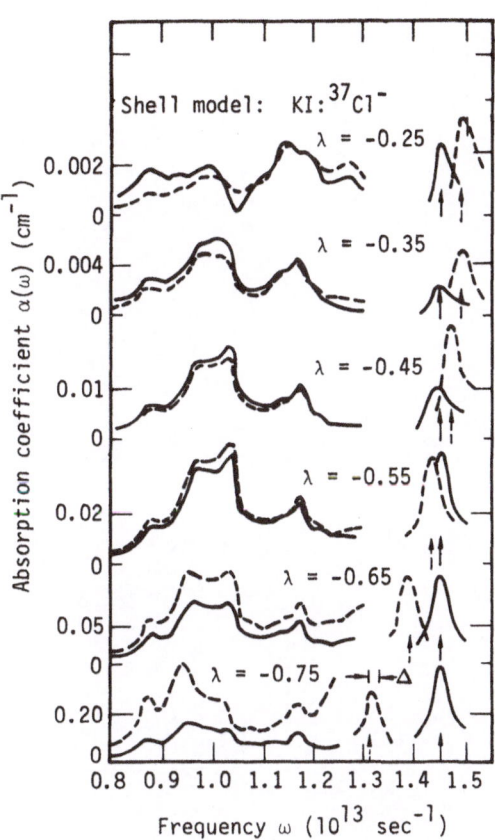

curves are those for which λ' has been adjusted to reproduce the observed gap mode frequency while the broken curves correspond to $\lambda' = 0$. Continuous and broken arrows indicate the positions of the corresponding gap mode frequencies. The parameter Δ is a measure of the width of the theoretical gap mode absorption peak.

V

Two-Phonon Data

9. Interpretation of the Infrared Spectra of Perfect Alkali Halides

In our earlier discussion of the interpretation of the infrared-absorption spectra of perfect alkali halides, we made use of the simplest possible classical treatment. The principal discrepancy between the absorption spectra calculated in this way and those observed experimentally lies in the existence of subsidiary structure in the observed spectra. In this section we wish to describe briefly how this can be explained. Since extensive discussions of the problem have been given by a number of authors,[49,50,148–150] we shall give a semiquantitative discussion of the absorption mechanism and show that the interpretation of the observed side-band structure can be reduced to the computation of certain functions of the eigen-frequencies and eigenvectors of the lattice.

There are two possible mechanisms that can cause the observed side-band structure. The first process is the more complicated and is the one associated with lattice anharmonicity. Although we are largely excluding any detailed consideration of anharmonic effects, we must consider this specific effect if we wish to understand the infrared-absorption properties in sufficient detail. This is because the anharmonic terms in the lattice potential energy are the origin of the damping that we introduced in Section 4 on an *ad hoc* basis.

A transverse optical phonon whose wave vector is approximately zero will exist for a finite period of time, after which it will decay into other phonons. The basic selection rule for such processes due to the cubic term in the lattice potential energy implies decay of the initial transverse optical phonon into two phonons having equal and opposite wave vectors. One can see this explicitly by writing out the third-order term in the potential:

$$U_3 = \tfrac{1}{6} \sum_{\substack{ll'l'' \\ kk'k'' \\ \alpha\beta\gamma}} \left[\partial^3 U \Big/ \partial\xi_\alpha\binom{l}{k}\, \partial\xi_\beta\binom{l'}{k'}\, \partial\xi_\gamma\binom{l''}{k''} \right]_0 \xi_\alpha\binom{l}{k} \xi_\beta\binom{l'}{k'} \xi_\gamma\binom{l''}{k''}$$

(9.1)

The direct-space displacements are then Fourier transformed and yield the final expression in terms of the normal coordinates $Q\binom{q}{j}$:

$$U_3 = (6N^{1/2})^{-1} \sum_{\substack{qq'q'' \\ jj'j''}} U\binom{qq'q''}{j\,j'\,j''} Q\binom{q}{j} Q\binom{q'}{j'} Q\binom{q''}{j''} \Delta(\mathbf{q}+\mathbf{q}'+\mathbf{q}'')$$

where $\Delta(\mathbf{q}) = 0$ unless $\mathbf{q}$ is a reciprocal lattice vector, in which case $\Delta(\mathbf{q}) = 1$, and

$$U\binom{qq'q''}{j\,j'\,j''} = \sum_{\substack{k\alpha \\ l'k'\beta \\ l''k''\gamma}} \left[\partial^3 U \Big/ \partial\xi_\alpha\binom{l}{k}\, \partial\xi_\beta\binom{l'}{k'}\, \partial\xi_\gamma\binom{l''}{k''} \right]_0$$

$$\times \left[e_\alpha\!\left(k\Big|\genfrac{}{}{0pt}{}{\mathbf{q}}{j}\right) e_\beta\!\left(k'\Big|\genfrac{}{}{0pt}{}{\mathbf{q}'}{j'}\right) e_\gamma\!\left(k''\Big|\genfrac{}{}{0pt}{}{\mathbf{q}''}{j''}\right) \Big/ (m_k m_{k'} m_{k''})^{1/2} \right]$$

$$\times \exp\left\{ 2\pi i \left[\mathbf{q}' \cdot \mathbf{r}\binom{l'}{k'} + \mathbf{q}'' \cdot \mathbf{r}\binom{l''}{k''} \right] \right\} \Big\}$$

(9.2)

Expressing the normal coordinates Q in terms of creation and destruction operators, it can be shown that the damping constant[148] for the mode $\binom{q}{j}$ measured at a frequency Ω is given by

$$\Gamma\left(\genfrac{}{}{0pt}{}{q}{j}\bigg|\Omega\right) \propto \sum_{\substack{q'q'' \\ j'j''}} \left|U\left(\genfrac{}{}{0pt}{}{q\,q'\,q''}{j\,j'\,j''}\right)\right|^2 \frac{\Delta(q+q'+q'')}{\omega\binom{q'}{j'}\omega\binom{q''}{j''}}\left\{\left[n\binom{q'}{j'}+n\binom{q''}{j''}+1\right]\right.$$

$$\times\delta\left[\Omega-\omega\binom{q'}{j'}-\omega\binom{q''}{j''}\right]$$

$$\left.+\left|n\binom{q'}{j'}-n\binom{q''}{j''}\right|\delta\left[\Omega-\left|\omega\binom{q'}{j'}-\omega\binom{q''}{j''}\right|\right]\right\} \qquad (9.3)$$

and when $q = 0$, we find that

$$q' = -q''$$

This damping constant is frequency dependent, and its fine structure is responsible for the corresponding detail in the infrared-absorption coefficient.

To proceed further, some approximation must be employed to describe the lattice anharmonicity. The simplest model, used by Cowley,[148] assumes that anharmonic effects are significant only for the short-range component of the interaction between first neighbors. We have simplified this further and assumed that the third derivative of this potential is dominant. Within this approximation the third-order anharmonic part of the potential is determined by a single constant. This is supported by the work of Eldridge and Howard.[151,152]

Thus, to examine the frequency dependence of the damping constant for a given rock-salt structure material, we must compute the following quantity for $j \equiv$ t.o., the transverse optical branch:

$$\Gamma\left(\begin{matrix} 0 \\ \text{t.o.} \end{matrix}\middle|\Omega\right) = \lim_{\substack{N\to\infty \\ \Delta\Omega\to 0}} \frac{A}{N\Delta\Omega} \int_{\Omega}^{\Omega+\Delta\Omega} d\Omega' \sum_{\substack{q \\ jj'}} \left\{ \left[e_x\left(1\middle|\begin{matrix} \mathbf{q} \\ j \end{matrix}\right) e_x\left(2\middle|\begin{matrix} -\mathbf{q} \\ j' \end{matrix}\right)\right.\right.$$

$$\left.\left. - e_x\left(2\middle|\begin{matrix} \mathbf{q} \\ j \end{matrix}\right) e_x\left(1\middle|\begin{matrix} -\mathbf{q} \\ j' \end{matrix}\right) \right] \sin 2\pi q_x r_0 \right\}^2$$

$$\times \left[m_1 m_2 \omega\left(\begin{matrix} \mathbf{q} \\ j \end{matrix}\right) \omega\left(\begin{matrix} -\mathbf{q} \\ j' \end{matrix}\right) \right]^{-1} \left\{ \left[n\left(\begin{matrix} \mathbf{q} \\ j \end{matrix}\right) + n\left(\begin{matrix} -\mathbf{q} \\ j' \end{matrix}\right) + 1 \right]\right.$$

$$\times \delta\left[\Omega' - \omega\left(\begin{matrix} \mathbf{q} \\ j \end{matrix}\right) - \omega\left(\begin{matrix} -\mathbf{q} \\ j' \end{matrix}\right) \right]$$

$$\left. + \left| n\left(\begin{matrix} \mathbf{q} \\ j \end{matrix}\right) - n\left(\begin{matrix} -\mathbf{q} \\ j' \end{matrix}\right) \right| \delta\left[\Omega' - \left| \omega\left(\begin{matrix} \mathbf{q} \\ j \end{matrix}\right) - \omega\left(\begin{matrix} -\mathbf{q} \\ j' \end{matrix}\right) \right| \right] \right\}$$

$$\tag{9.4}$$

We have done this for all these alkali halides, and as an illustration we show the results for the sodium halide sequence in Figs. 35–38. In no case have we attempted to make any estimate of the absolute magnitude of the constant A in Eq. (9.4); thus our values of the damping constants are in arbitrary units, our only concern being to display their structure.

The other distinct mechanism that can produce substructure in the infrared-absorption spectrum is the second-order dipole moment. The existence of this mechanism was first postulated by Lax and Burstein,[153] when they examined in detail the coupling between an external electromagnetic field and the normal modes of a crystal lattice. For ionic crystals one normally considers only the linear term, an approach that would be rigorous in the absence of ionic deformation. However, to allow for the possibility of the deformation of the electronic charge clouds about the ions by lattice

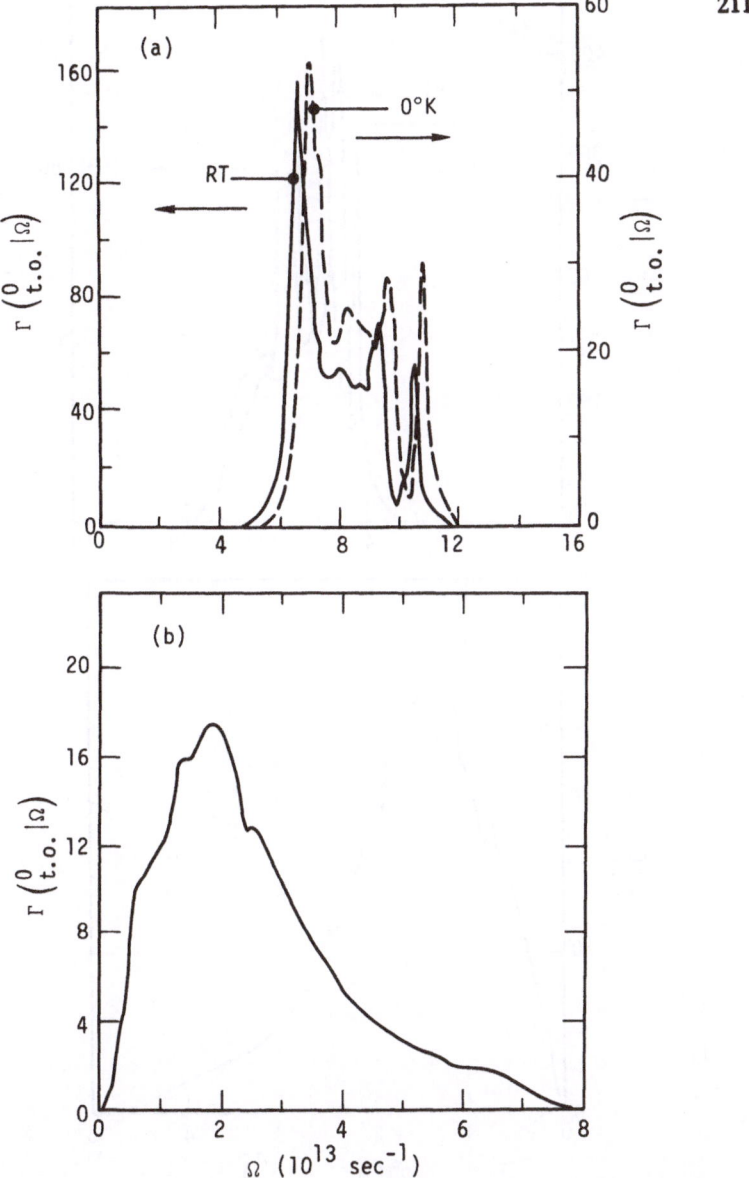

FIG. 35. Frequency dependence of the damping constant for NaF calculated according to Eq. (9.4) for (a) combination terms (first δ function) and (b) difference terms (second δ function). Solid and broken curves denote room- and low-temperature input data, respectively. The model used is the DD3N model (deformation dipole with short-range forces extended to include closest-neighbor negative ions).

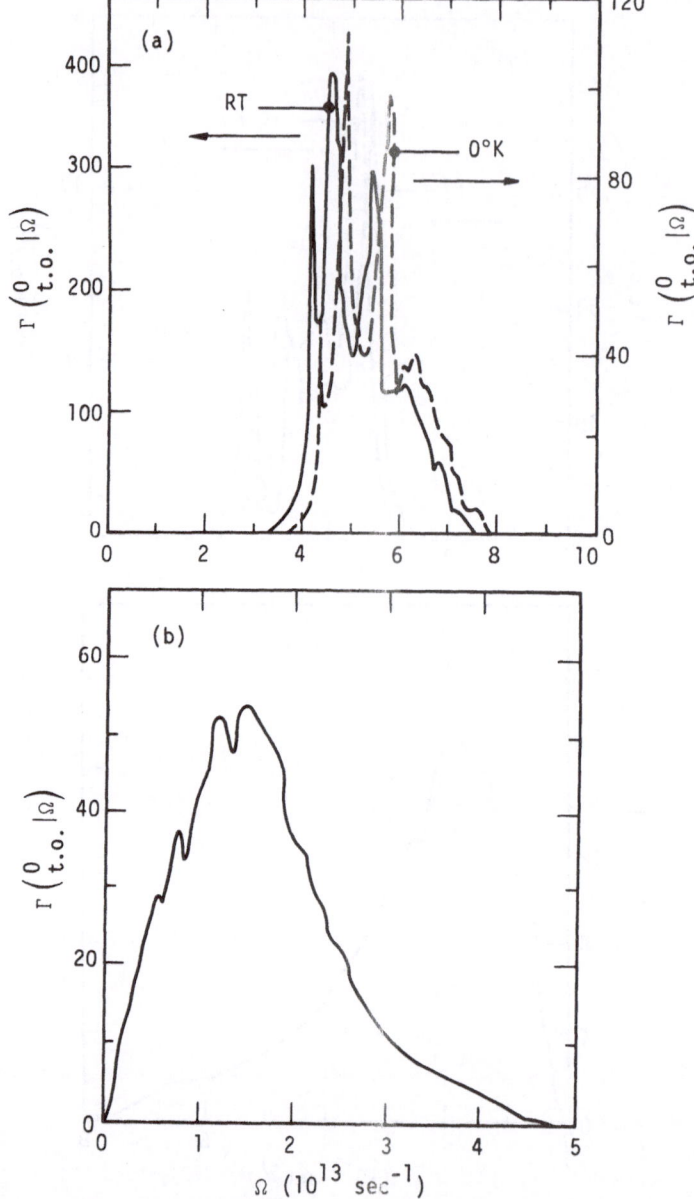

FIG. 36. Frequency dependence of the damping constant for NaCl calculated according to Eq. (9.4) for (a) combination terms (first δ function) and (b) difference terms (second δ function). Solid and broken curves denote room- and low-temperature input data, respectively. The model used is the DD3N model (deformation dipole with short-range forces extended to include closest-neighbor negative ions).

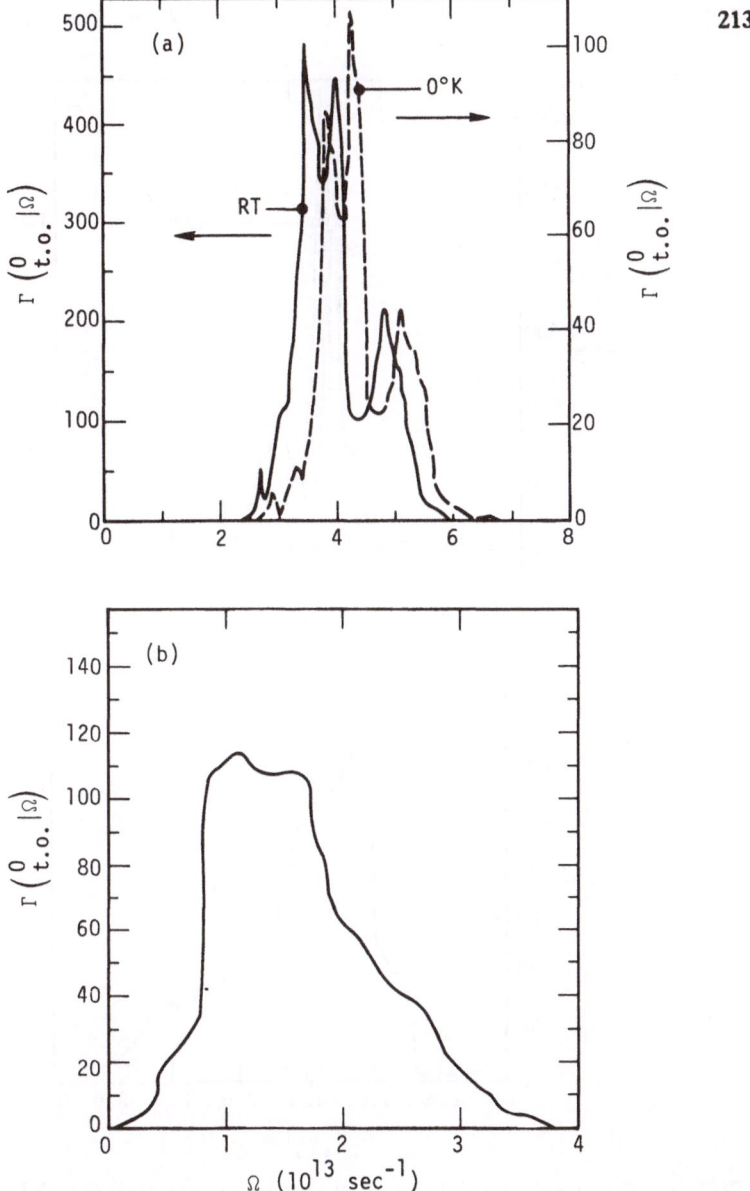

FIG. 37. Frequency dependence of the damping constant for NaBr calculated according to Eq. (9.4) for (a) combination terms (first δ function) and (b) difference terms (second δ function). Solid and broken curves denote room- and low-temperature input data, respectively. The model used is the DD3N model (deformation dipole with short-range forces extended to include closest-neighbor negative ions).

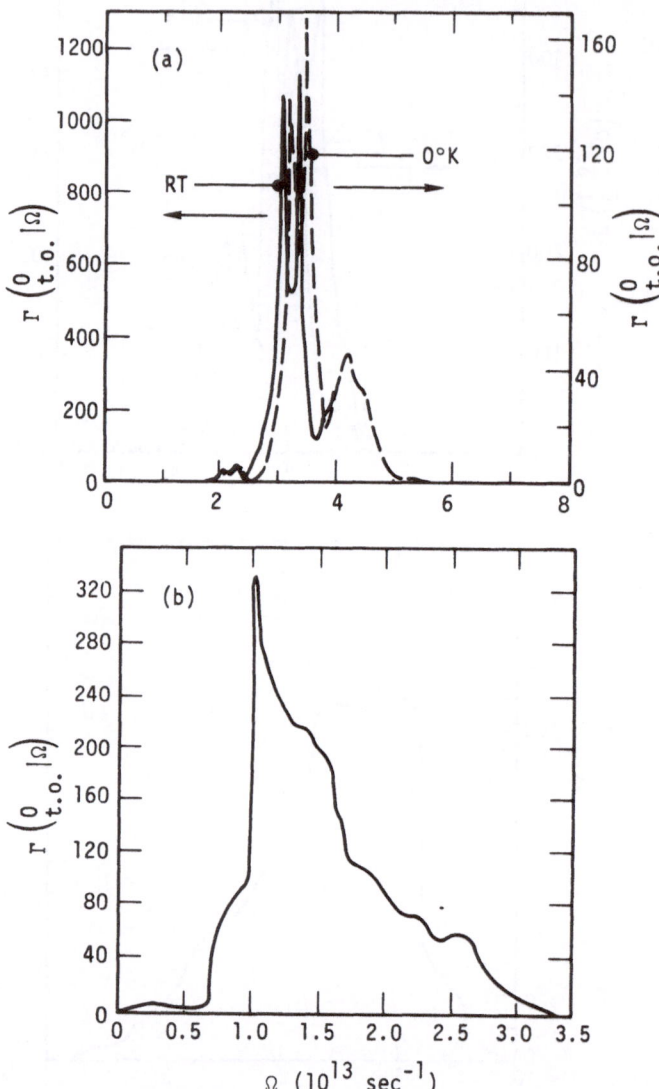

FIG. 38. Frequency dependence of the damping constant for NaI calculated according to Eq. (9.4) for (a) combination terms (first δ function) and (b) difference terms (second δ function). Solid and broken curves denote room- and low-temperature input data, respectively. The model used is the DD3N model (deformation dipole with short-range forces extended to include closest-neighbor negative ions).

distortion is to introduce the possibility of a coupling to the external electromagnetic field that is finite to all orders in the displacements. The most important additional term is the one that is second-order in the displacements. Again, Fourier transformation of the displacements yields for this second-order term

$$M_\alpha^{(2)} = \tfrac{1}{2} \sum_{\substack{qj \\ \beta\gamma}} \sum_{l'kk'} (m_k m_{k'})^{-1/2} \left[\partial^2 M_\alpha \Big/ \partial \xi_\beta \binom{0}{k} \partial \xi_\gamma \binom{l'}{k'} \right]_0$$

$$\times e_\beta \left(k \Big| \begin{matrix} \mathbf{q} \\ j \end{matrix} \right) e_\gamma \left(k' \Big| \begin{matrix} -\mathbf{q} \\ j' \end{matrix} \right) \exp \left\{ -2\pi i \mathbf{q} \cdot \left[\mathbf{r}\binom{l'}{k'} \right. \right.$$

$$\left. \left. - \mathbf{r}\binom{0}{k} \right] \right\} Q\binom{\mathbf{q}}{j} Q\binom{-\mathbf{q}}{j'} \tag{9.5}$$

There is therefore the possibility of a new absorption process, whose cross section is given by

$$\sigma(\Omega) \propto \lim_{\substack{N\to\infty \\ \Delta\Omega\to 0}} \frac{1}{N\Delta\Omega} \int_\Omega^{\Omega+\Delta\Omega} d\Omega' \sum_{qjj'} \left[\partial^2 M_\alpha^{(2)} \Big/ \partial Q\binom{\mathbf{q}}{j} \partial Q\binom{-\mathbf{q}}{j'} \right]^2$$

$$\times \left[\omega\binom{\mathbf{q}}{j} \omega\binom{-\mathbf{q}}{j'} \right]^{-1}$$

$$\times \left\{ \left[n\binom{\mathbf{q}}{j} + n\binom{-\mathbf{q}}{j'} + 1 \right] \delta \left[\Omega' - \omega\binom{\mathbf{q}}{j} - \omega\binom{-\mathbf{q}}{j'} \right] \right.$$

$$\left. + \left| n\binom{\mathbf{q}}{j} - n\binom{-\mathbf{q}}{j'} \right| \delta \left[\Omega' - \left| \omega\binom{\mathbf{q}}{j} - \omega\binom{-\mathbf{q}}{j'} \right| \right] \right\} \tag{9.6}$$

For ionic crystals, there is cross-coupling between two-phonon creation by lattice anharmonicity and by second-order dipole-moment effects.[154] This leads to complications that are not present in the case of homopolar crystals, and the computation of the optical constants is an involved process.[155,156] However, the basic lattice-dynamical information is contained in the functions defined by Eqs. (9.4) and (9.6), and it is to these that we confine our attention.

(It should be added at this point that there is a further possible complication whose existence was pointed out by Szigeti.[155,156] He showed that large anharmonicity is to be expected in materials that have large second-order dipole moments. This is because there is a direct coupling between the first-order dipole moment associated with a given lattice mode and the second-order dipole moment associated with any other pair of lattice modes. This coupling produces a third-order anharmonic interaction between the modes.)

Fortunately, if it is assumed that the origin of the second-order dipole moment is entirely due to first-neighbor overlap, then the quantity σ in Eq. (9.6) has the same form as Γ in Eq. (9.4), and the two need not be computed separately. Correlating the structure of both functions with that observed in the measured spectra can thus be attempted.

It can be shown that the computed damping constants for the sodium halide sequence (Figs. 35–38) automatically satisfy the group-theory selection rules given by Loudon.[157] (Since our model is oversimplified, we have additional restrictions on the allowed two-phonon combinations—restrictions that are not required by group theory.)

Since these rules are fairly stringent, they tend to suppress the fine structure that one obtains from simple combined densities of states computed according to the following prescriptions:

$$N_2^+ (\Omega) = \lim_{\substack{N \to \infty \\ \Delta\Omega \to 0}} (N\Delta\Omega)^{-1} \sum_{\substack{\mathbf{q} \\ jj'}} \int_{\Omega}^{\Delta + \Delta\Omega} \left[n\binom{\mathbf{q}}{j} + n\binom{-\mathbf{q}}{j'} + 1 \right]$$

$$\times \delta\left[\Omega' - \omega\binom{\mathbf{q}}{j} - \omega\binom{-\mathbf{q}}{j'} \right] d\Omega' \qquad (9.7a)$$

and

$$N_2^- (\Omega) = \lim_{\substack{N \to \infty \\ \Delta\Omega \to 0}} (N\Delta\Omega)^{-1} \sum_{\substack{\mathbf{q} \\ \mathit{jj'}}} \int_{\Omega}^{\Omega+\Delta\Omega} \left| n\binom{\mathbf{q}}{j} - n\binom{-\mathbf{q}}{j'} \right|$$

$$\times \delta \left[\Omega' - \left| \omega\binom{\mathbf{q}}{j} - \omega\binom{-\mathbf{q}}{j'} \right| \right] d\Omega' \qquad (9.7b)$$

The resultant curves for the sodium halide sequence are shown in Figs. 39–42. There is considerable correlation between the peaks in these spectra and the observed side bands in the infrared-absorption spectra.[158]

It is reasonable to expect that side-band spectra would be rather featureless. This is certainly the case for the alkali halide two-phonon side-band spectra, as compared with Johnson's[159] measurements on silicon and germanium. The reasons for this are twofold: first, in the case of the alkali halides, the selection rules are more stringent; second, the shapes of the alkali halide dispersion curves are very different from those of the group IV elements. In particular, the alkali halides do not have any nearly flat phonon branches such as those that are present for the group IV materials. For the reasons we have just given, it is our feeling that the frequency-spectra information that can be extracted by analyzing infrared-absorption data is limited.

Finally, we should mention that extensive studies have been made of the infrared-absorption spectra associated with substitutional impurities in the alkali halides. In particular, we would cite the work on local-mode absorption spectra due to U centers[160] (substitutional H^- ions). These have a definite side-band structure whose origin, discussed by Bilz et al.,[161] is the simultaneous creation of a local-mode phonon and a lattice phonon. It appears that this structure is due to the fact that the lattice potential energy contains anharmonic terms that are bilinear in the local-mode amplitude and

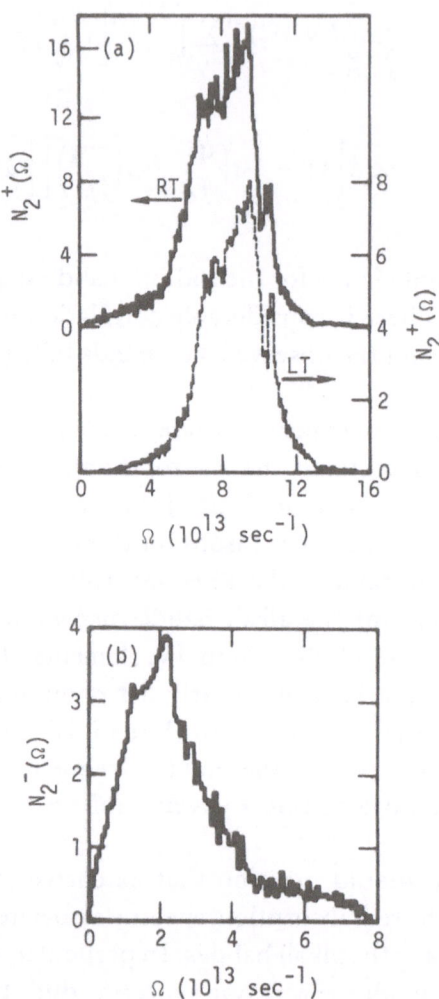

FIG. 39. Combined densities of states for NaF calculated according to Eqs. (9.7a) and (9.7b) for (a) combination terms and (b) difference terms, respectively. Solid and broken curves denote room- and low-temperature results, respectively (using appropriate input data). The model used is the DD3N model (deformation dipole with short-range forces extended to include closest-neighbor negative ions).

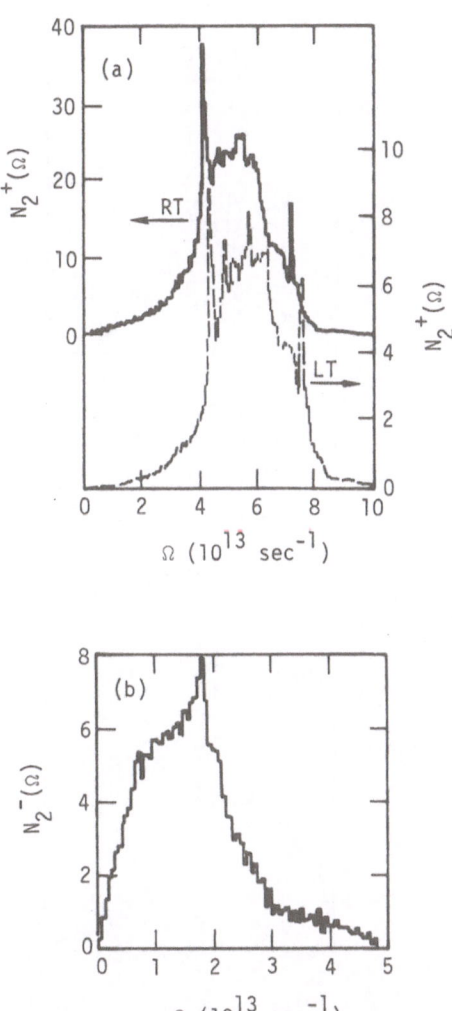

FIG. 40. Combined densities of states for NaCl calculated according to Eqs. (9.7a) and (9.7b) for (a) combination terms and (b) difference terms, respectively. Solid and broken curves denote room- and low-temperature results, respectively (using appropriate input data). The model used is the DD3N model (deformation dipole with short-range forces extended to include closest-neighbor negative ions).

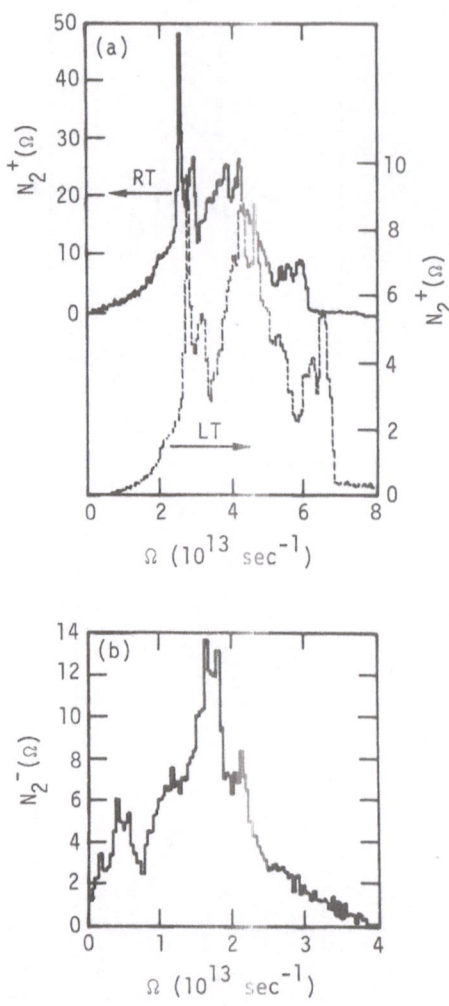

FIG. 41. Combined densities of states for NaBr calculated according to Eqs. (9.7a) and (9.7b) for (a) combination terms and (b) difference terms, respectively. Solid and broken curves denote room- and low-temperature results, respectively (using appropriate input data). The model used is the DD3N model (deformation dipole with short-range forces extended to include closest-neighbor negative ions).

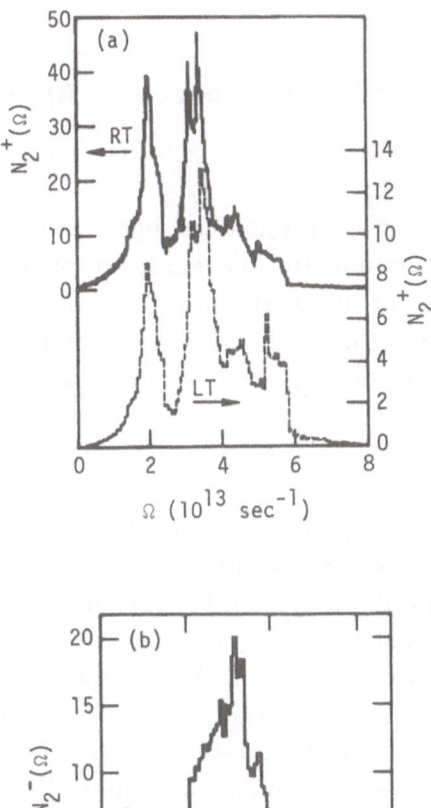

FIG. 42. Combined densities of states for NaI calculated according to Eqs. (9.7a) and (9.7b) for (a) combination terms and (b) difference terms, respectively. Solid and broken curves denote room- and low-temperature results, respectively (using appropriate input data). The model used is the DD3N model (deformation dipole with short-range forces extended to include closest-neighbor negative ions).

linear in the lattice-mode amplitude. For an extensive discussion of these impurity effects, we refer the reader to Klein's review.[162]

10. Second-Order Raman Spectra of Alkali Halide Crystals

With the advent of laser sources, light scattering has become a very effective technique for the study of phonons. Although a considerable amount of work had been done previously with mercury-arc sources,‡ in most cases the results so obtained were not particularly reliable and, in general, no polarization information was obtained.

Raman scattering is the inelastic scattering of photons by phonons or any other elementary excitation of the system under investigation. Here we shall confine ourselves to Raman scattering by phonons. It should be stated at the outset that the theory is still very incomplete. The ideal theory would be capable of predicting both the shape and the intensity of the observed spectra from a detailed quantum-mechanical calculation. At present the calculations that have come closest to this ideal are those of Loudon[164] on the first-order Raman-scattering cross section for polar and nonpolar semiconductors. However, similar calculations for the second-order spectra have yet to be made and are likely to be extremely complicated.

Since our interest is centered on predicting the shapes of the second-order spectra, we have to fall back on the phenomenological approach originally due to Born and Bradburn[165] and described in general terms by Born and Huang.[166] Thus the Raman polarizability tensor is characterized by a number of independent parameters,

---

‡ A comprehensive list of studies made with mercury-arc sources is given by Krishnan.[163]

which are the second derivatives of the crystal polarizability with respect to pairs of nuclear coordinates. It is then necessary to reduce the number of these parameters by such means as symmetry considerations, until meaningful predictions about the shape of the spectrum can be made. It is not possible by this approach to make any prediction of the absolute magnitude of the scattering. For the present, the shape of the spectrum is much more interesting, and it depends only on the relative magnitudes of the various parameters.

Recent experimental work[167-170] indicates that it is possible to obtain an understanding of the Raman spectra of the rock-salt-structure alkali halides without introducing an excessively large number of parameters. The important point about the Raman spectra is that they contain considerably more information than do the second-order infrared spectra. In particular, it is possible to extract important information from the relative strengths of the polarized and depolarized spectra. As discussed in detail later, it is this type of information that has made possible important progress in understanding the Raman spectra of alkali halides.

The phenomenological equation for the second-order Raman polarizability tensor has the following form:

$$P_{\alpha\beta}^{(2)} = \frac{1}{2}\sum_{\substack{ll' \\ kk' \\ \gamma\lambda}} \left[\partial^2 P_{\alpha\beta}\Big/\partial\xi_\gamma\binom{l}{k}\,\partial\xi_\lambda\binom{l'}{k'}\right]_0 \xi_\gamma\binom{l}{k}\xi_\lambda\binom{l'}{k'} \quad (10.1)$$

This is obtained by expanding the electronic polarizability of the crystal lattice as a power series in the nuclear displacements. The dependence on the nuclear displacements is a second-order nonlinear effect, and consequently the observed Raman-scattering cross-sections for alkali halide crystals are very small. A formal perturbation-theory calculation can be shown to give the coefficients

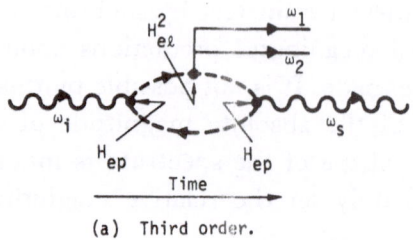

(a)   Third order.

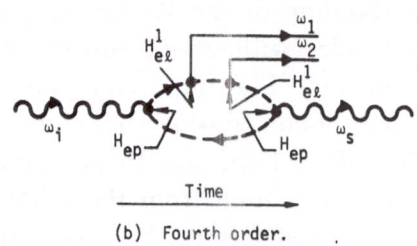

(b)   Fourth order.

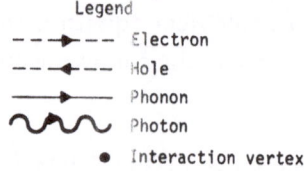

FIG. 43. Feynman-like diagrams showing possible third- and fourth-order processes giving rise to second-order Raman scattering. H_{ep} corresponds to the linear electron–photon coupling, while H_{el}^1 and H_{el}^2 correspond to linear and quadratic electron–phonon coupling, respectively. *Note*: Second-order scattering is generally dominated by the quadratic terms in the electron–phonon interaction, which make it possible as a third-order process.

in this expansion.[171] Quadratic terms in the lattice displacements can be shown to arise from both third- and fourth-order perturbation-theoretic calculations. Typical associated Feynman diagrams are shown in Fig. 43. For first-order scattering, linear electron–phonon coupling alone need be considered, but is generally not adequate for calculating the second-order Raman-scattering cross section.

The polarizability tensor that one would calculate by this procedure can be put into the phenomenological form

$$\frac{1}{2}\sum_{\substack{\mathbf{q}\\ jj'}} P^{(2)}_{\alpha\beta}\left(\begin{matrix}\mathbf{q}\\ j\end{matrix}\middle|\begin{matrix}-\mathbf{q}\\ j'\end{matrix}\right) Q\left(\begin{matrix}\mathbf{q}\\ j\end{matrix}\right) Q\left(\begin{matrix}-\mathbf{q}\\ j'\end{matrix}\right) = \frac{1}{2}\sum_{\substack{\mathbf{q}k\gamma\\ lk'\lambda\\ jj'}} \left[P_{\alpha\beta,\gamma\lambda}\left(\begin{matrix}l\\ kk'\end{matrix}\right)\middle/ (m_k m_{k'})^{1/2}\right]$$

$$\times e_\gamma\left(k\middle|\begin{matrix}\mathbf{q}\\ j\end{matrix}\right) e_\lambda\left(k'\middle|\begin{matrix}-\mathbf{q}\\ j'\end{matrix}\right)$$

$$\times \exp\left\{-2\pi i \mathbf{q}\cdot\left[\mathbf{r}\left(\begin{matrix}l\\ k\end{matrix}\right)\right.\right.$$

$$\left.\left.-\mathbf{r}\left(\begin{matrix}0\\ k\end{matrix}\right)\right]\right\} Q\left(\begin{matrix}\mathbf{q}\\ j\end{matrix}\right) Q\left(\begin{matrix}-\mathbf{q}\\ j'\end{matrix}\right) \qquad (10.2)$$

where $P_{\alpha\beta,\gamma\lambda}\left(\begin{smallmatrix}l\\ kk'\end{smallmatrix}\right)$ is an abbreviation for the coefficient in square brackets in Eq. (10.1).

As already stated, the object of the present procedure is to reduce the number of parameters in Eq. (10.2) as far as possible. An initial reduction in the number of parameters can be achieved by assuming that the polarizability tensor is affected by relative displacements of first neighbors only. We then obtain the following equations for $P^{(2)}_{zz}\left(\begin{smallmatrix}\mathbf{q}\\ j\end{smallmatrix}\middle|\begin{smallmatrix}-\mathbf{q}\\ j'\end{smallmatrix}\right)$ and $P^{(2)}_{yz}\left(\begin{smallmatrix}\mathbf{q}\\ j\end{smallmatrix}\middle|\begin{smallmatrix}-\mathbf{q}\\ j'\end{smallmatrix}\right)$ (all other components can be obtained by permuting x, y, and z):

$$P_{zz}^{(2)}\left(\begin{matrix}\mathbf{q}\\j\end{matrix}\Big|\begin{matrix}-\mathbf{q}\\j'\end{matrix}\right) = \Bigg\{(a+2b)\bigg[\sigma_z\left(k\Big|\begin{matrix}\mathbf{q}\\j\end{matrix}\right)\sigma_z\left(k\Big|\begin{matrix}-\mathbf{q}\\j'\end{matrix}\right)$$

$$+\,\sigma_z\left(k'\Big|\begin{matrix}\mathbf{q}\\j\end{matrix}\right)\sigma_z\left(k'\Big|\begin{matrix}-\mathbf{q}\\j'\end{matrix}\right)\bigg]$$

$$+\,(c+d_1+d_2)\bigg[\sigma_x\left(k\Big|\begin{matrix}\mathbf{q}\\j\end{matrix}\right)\sigma_x\left(k\Big|\begin{matrix}-\mathbf{q}\\j'\end{matrix}\right)$$

$$+\,\sigma_x\left(k'\Big|\begin{matrix}\mathbf{q}\\j\end{matrix}\right)\sigma_x\left(k'\Big|\begin{matrix}-\mathbf{q}\\j'\end{matrix}\right)$$

$$+\,\sigma_y\left(k\Big|\begin{matrix}\mathbf{q}\\j\end{matrix}\right)\sigma_y\left(k\Big|\begin{matrix}-\mathbf{q}\\j'\end{matrix}\right)+\sigma_y\left(k'\Big|\begin{matrix}\mathbf{q}\\j\end{matrix}\right)\sigma_y\left(k'\Big|\begin{matrix}-\mathbf{q}\\j'\end{matrix}\right)\bigg]\Bigg\}$$

$$-\Bigg\{\bigg[\sigma_z\left(k'\Big|\begin{matrix}\mathbf{q}\\j\end{matrix}\right)\sigma_z\left(k\Big|\begin{matrix}-\mathbf{q}\\j'\end{matrix}\right)+\sigma_z\left(k\Big|\begin{matrix}\mathbf{q}\\j\end{matrix}\right)\sigma_z\left(k'\Big|\begin{matrix}-\mathbf{q}\\j'\end{matrix}\right)\bigg]$$

$$\times\,[a\cos 2\pi q_z r_0 + b(\cos 2\pi q_x r_0 + \cos 2\pi q_y r_0)]$$

$$+\bigg[\sigma_x\left(k'\Big|\begin{matrix}\mathbf{q}\\j\end{matrix}\right)\sigma_x\left(k\Big|\begin{matrix}-\mathbf{q}\\j'\end{matrix}\right)+\sigma_x\left(k\Big|\begin{matrix}\mathbf{q}\\j\end{matrix}\right)\sigma_x\left(k'\Big|\begin{matrix}-\mathbf{q}\\j'\end{matrix}\right)\bigg]$$

$$\times\,(d_2\cos 2\pi q_x r_0 + c\cos 2\pi q_y r_0 + d_1\cos 2\pi q_z r_0)$$

$$+\bigg[\sigma_y\left(k'\Big|\begin{matrix}\mathbf{q}\\j\end{matrix}\right)\sigma_y\left(k\Big|\begin{matrix}-\mathbf{q}\\j'\end{matrix}\right)+\sigma_y\left(k\Big|\begin{matrix}\mathbf{q}\\j\end{matrix}\right)\sigma_y\left(k'\Big|\begin{matrix}-\mathbf{q}\\j'\end{matrix}\right)\bigg]$$

$$\times\,(c\cos 2\pi q_x r_0 + d_2\cos 2\pi q_y r_0 + d_1\cos 2\pi q_z r_0)\Bigg\}$$

$$\tag{10.3a}$$

$$P_{yz}^{(2)}\left(\begin{matrix}\mathbf{q}\\j\end{matrix}\Big|\begin{matrix}-\mathbf{q}\\j'\end{matrix}\right) = (e+f+g)\bigg[\sigma_y\left(k\Big|\begin{matrix}\mathbf{q}\\j\end{matrix}\right)\sigma_z\left(k\Big|\begin{matrix}-\mathbf{q}\\j'\end{matrix}\right)$$ *(equation continues on next page)*

$$\tag{10.3b}$$

(equation (10.3b) continued)

$$+ \sigma_z\left(k \left| \begin{matrix} \mathbf{q} \\ j \end{matrix}\right.\right) \sigma_y\left(k \left| \begin{matrix} -\mathbf{q} \\ j' \end{matrix}\right.\right)$$

$$+ \sigma_y\left(k' \left| \begin{matrix} \mathbf{q} \\ j \end{matrix}\right.\right) \sigma_z\left(k' \left| \begin{matrix} -\mathbf{q} \\ j' \end{matrix}\right.\right) + \sigma_z\left(k' \left| \begin{matrix} \mathbf{q} \\ j \end{matrix}\right.\right) \sigma_y\left(k' \left| \begin{matrix} -\mathbf{q} \\ j' \end{matrix}\right.\right) \Big]$$

$$- \left\{ \left[\sigma_y\left(k' \left| \begin{matrix} \mathbf{q} \\ j \end{matrix}\right.\right) \sigma_z\left(k \left| \begin{matrix} -\mathbf{q} \\ j' \end{matrix}\right.\right) + \sigma_y\left(k \left| \begin{matrix} \mathbf{q} \\ j \end{matrix}\right.\right) \sigma_z\left(k' \left| \begin{matrix} -\mathbf{q} \\ j' \end{matrix}\right.\right) \right] \right.$$

$$\times (e \cos 2\pi q_x r_0 + f \cos 2\pi q_y r_0 + g \cos 2\pi q_z r_0)$$

$$+ \left[\sigma_z\left(k' \left| \begin{matrix} \mathbf{q} \\ j \end{matrix}\right.\right) \sigma_y\left(k \left| \begin{matrix} -\mathbf{q} \\ j' \end{matrix}\right.\right) + \sigma_z\left(k \left| \begin{matrix} \mathbf{q} \\ j \end{matrix}\right.\right) \sigma_y\left(k' \left| \begin{matrix} -\mathbf{q} \\ j' \end{matrix}\right.\right) \right]$$

$$\times \left. (e \cos 2\pi q_x r_0 + f \cos 2\pi q_z r_0 + g \cos 2\pi q_y r_0) \right\}$$

The eight constants $a, b, \ldots, g$ are the eight independent polarizability derivatives; as before r_0 is the nearest-neighbor distance; the $\sigma(k|\begin{smallmatrix} \mathbf{q} \\ j \end{smallmatrix})$ terms are the eigenvectors $e_\alpha(k|\begin{smallmatrix} \mathbf{q} \\ j \end{smallmatrix})$ of the dynamical matrix divided by the square roots of the appropriate masses; and $k = 1$, $k' = 2$. Obviously, even allowing for the fact that we are interested only in the relative intensities and thus have only seven unknowns, we have too many parameters.

In order to reduce the number of these parameters further, we assume that the polarizability tensor depends only on the *relative* separation of the two neighbors in a given bond. We then find that the parameters are reduced in number since

$$b = c = d_1 \quad \text{and} \quad e = f = g = 0$$

Consequently we now have three disposable parameters all told— and only two if we are interested in relative intensities.

To complete the reduction, we assume that either

$$c = d_2 = 0 \qquad \qquad \text{(variation 1)}$$

or

$$a = d_2, \qquad c = (-2\rho/r_0)a \qquad \text{(variation 2)}$$

where ρ is the Born–Mayer screening radius. These two alternatives we believe to represent two extremes. In this review we shall restrict ourselves to considering a specific scattering geometry. We consider the situation in which the incident beam lies along the [010] direction and is polarized along the [001] direction. We now consider the intensities of the polarized and depolarized scattered radiations viewed along the [100] direction. These are, respectively, given by

$$I_\parallel(\Omega) = \lim_{\substack{N\to\infty \\ \Delta\Omega\to 0}} \frac{B}{N\Delta\Omega} \sum_{\mathbf{q}jj'} \int_\Omega^{\Omega+\Delta\Omega} d\Omega' \left\{ \left[n\binom{\mathbf{q}}{j} + 1 \right]\left[n\binom{-\mathbf{q}}{j'} + 1 \right] \right.$$

$$\times \delta\left[\Omega' - \omega\binom{\mathbf{q}}{j} - \omega\binom{-\mathbf{q}}{j'} \right] + \left[n\binom{\mathbf{q}}{j} + 1 \right] n\binom{-\mathbf{q}}{j'}$$

$$\times \delta\left[\Omega' - \omega\binom{\mathbf{q}}{j} + \omega\binom{-\mathbf{q}}{j'} \right] + n\binom{\mathbf{q}}{j} n\binom{-\mathbf{q}}{j'}$$

$$\left. \times \delta\left[\Omega' + \omega\binom{\mathbf{q}}{j} + \omega\binom{-\mathbf{q}}{j'} \right] \right\}$$

$$\times \left\{ P_{zz}\binom{\mathbf{q}}{j}\bigg|\binom{-\mathbf{q}}{j'} \bigg/ \left[\omega\binom{\mathbf{q}}{j} \omega\binom{-\mathbf{q}}{j'} \right]^{1/2} \right\}^2 \qquad (10.4a)$$

and

$$I_\perp(\Omega) = \lim_{\substack{N\to\infty \\ \Delta\Omega\to 0}} \frac{B}{N\Delta\Omega} \sum_{\mathbf{q}jj'} \int_\Omega^{\Omega+\Delta\Omega} d\Omega' \left\{ \left[n\binom{\mathbf{q}}{j} + 1 \right]\left[n\binom{-\mathbf{q}}{j'} + 1 \right] \right.$$

(equation continues
on next page)   (10.4b)

(equation (10.4b) continued)

$$\times \delta\left[\Omega' - \omega\binom{\mathbf{q}}{j} - \omega\binom{-\mathbf{q}}{j''}\right] + \left[n\binom{\mathbf{q}}{j} + 1\right]n\binom{-\mathbf{q}}{j''}$$

$$\times \delta\left[\Omega' - \omega\binom{\mathbf{q}}{j} + \omega\binom{-\mathbf{q}}{j''}\right] + n\binom{\mathbf{q}}{j}n\binom{-\mathbf{q}}{j''}$$

$$\times \delta\left[\Omega' + \omega\binom{\mathbf{q}}{j} + \omega\binom{-\mathbf{q}}{j''}\right]\Bigg\}$$

$$\times \left\{P_{yz}\binom{\mathbf{q}}{j}\bigg|\begin{matrix}-\mathbf{q}\\j''\end{matrix}\right) \Bigg/ \left[\omega\binom{\mathbf{q}}{j}\omega\binom{-\mathbf{q}}{j''}\right]^{1/2}\right\}^2$$

With the foregoing assumptions, the depolarized component is zero for this geometry. This is supported to some extent by the results of experiments with sodium and potassium chlorides[167,168] and sodium fluoride,[169,170] which established that the depolarized component is generally an order of magnitude weaker than the polarized component. Thus one can see that polarization measurements have given very significant information about the components of the polarizability tensor.

In the case of cubic crystals, the Raman spectrum for any orientation and any set of relative polarizations of ingoing and outgoing radiation is uniquely determined by three independent components, denoted by $I(\Omega)_{\alpha\alpha,\alpha\alpha}$, $I(\Omega)_{\alpha\alpha,\beta\beta}$, and $I(\Omega)_{\alpha\beta,\alpha\beta}$, of the fourth-rank tensor $I(\Omega)_{\alpha\gamma,\beta\lambda}$:

$$I(\Omega)_{\alpha\gamma,\beta\lambda} = \lim_{\Delta\Omega \to 0} (\Delta\Omega)^{-1} \sum_{n'} (\langle n'|P^*_{\alpha\gamma}|n\rangle\langle n|P_{\beta\lambda}|n'\rangle)_{av} \quad (10.5)$$

$$(\Omega < \omega_0 + \omega_{nn'} < \Omega + \Delta\Omega)$$

Here ω_0 is the laser frequency and the ground and excited states of the lattice are denoted by $|n\rangle$ and $|n'\rangle$, respectively. The average is a thermal average over all possible states $|n\rangle$.

The prescription for calculating the scattered intensity for any arbitrary orientation of both the crystal and the polarization vectors of the incident and scattered light is given in a paper by Cowley.[172]

For the instrumental geometry described previously, one determines the two components

$$I(\Omega)_{zz,zz} = I_{\parallel}(\Omega) \quad \text{and} \quad I(\Omega)_{yz,yz} = I_{\perp}(\Omega)$$

Plots of $I_{\parallel}(\Omega)$ for variations 1 and 2 are shown in Figs. 44–47. To determine the third component, $I(\Omega)_{yy,zz}$, one needs to use some other experimental geometry, and, indeed, one does not determine this particular component by itself, but rather in combination with the other two. The determination of this component is of some importance since the two variations that we use differ drastically in the predicted form of $I(\Omega)_{\alpha\alpha,\beta\beta}$; specifically, $I(\Omega)_{\alpha\alpha,\beta\beta} = I(\Omega)_{\alpha\alpha,\alpha\alpha}$ for the second variation, whereas there is no simple relation for the first. To illustrate this point we present the results of calculations of $I_{yy,zz}$ in Figs. 44–47 for variation 1, since $I(\Omega)_{\alpha\alpha,\beta\beta} \neq I(\Omega)_{\alpha\alpha,\alpha\alpha}$ for this variation.

It is interesting to compare these results with those obtained by the simple combined density of states. The correlation between the two is variable. However, the structure of the combined density of states is less affected in Raman spectra than it is in infrared spectra, since the selection rules here are much weaker. Nonetheless, it is often extremely important to incorporate the polarizability tensor into the calculations in order to obtain even qualitative agreement between theory and experiment. When one does so using the models previously described, then, at least for sodium fluoride and sodium chloride, the agreement between theory and experiment is satisfactory. There is no obvious reason for this nearest-neighbor approximation's being so good, and a deeper under-standing of this result must wait on detailed quantum-mechanical calculations.

Bruce and Cowley[173] and Bruce[174] have recently made interesting calculations, with anharmonic shell models,[172] that throw some light on this point.

In the course of our past computations, in particular when computing the combined densities of states,[109] we have also shown the appropriate two-phonon dispersion curves since Van Hove singularities in the two-phonon density of states occur whenever

$$\left| \nabla_{\mathbf{q}} \left[\omega \begin{pmatrix} \mathbf{q} \\ j \end{pmatrix} \pm \omega \begin{pmatrix} -\mathbf{q} \\ j' \end{pmatrix} \right] \right| = 0$$

and this will be the case where these dispersion curves have zero slope. It should be noted that such a singularity can occur when

$$\nabla_{\mathbf{q}} \left[\omega \begin{pmatrix} \mathbf{q} \\ j \end{pmatrix} \right] = \mp \nabla_{\mathbf{q}} \left[\omega \begin{pmatrix} -\mathbf{q} \\ j' \end{pmatrix} \right]$$

Thus the total number of singularities in the two-phonon density of states is considerably larger than what would be obtained by pairing the points at which $|\nabla_{\mathbf{q}}[\omega(\begin{smallmatrix} \mathbf{q} \\ j \end{smallmatrix})]| = |\nabla_{\mathbf{q}}[\omega(\begin{smallmatrix} -\mathbf{q} \\ j' \end{smallmatrix})]| = 0$. Indeed, from the assignments that have been made we can see that it is usually impossible to make a complete and unambiguous critical-point assignment to the features of the two-phonon density of states.[109] This is characteristic of the alkali halide sequence of crystals. One can contrast this situation with the analysis[50] of the two-phonon infrared spectra of the group IV elements, where the singularities are much more pronounced and possibly fewer in number—a fact that reflects the less complex topology of the dispersion surfaces for these materials.

It would also appear that the computed difference spectra contain additional information about the elements of the polarizability tensor, and hence we have displayed these spectra separately. Consequently, experimental measurements of these spectra are of importance. It should be noted that the combined density of states

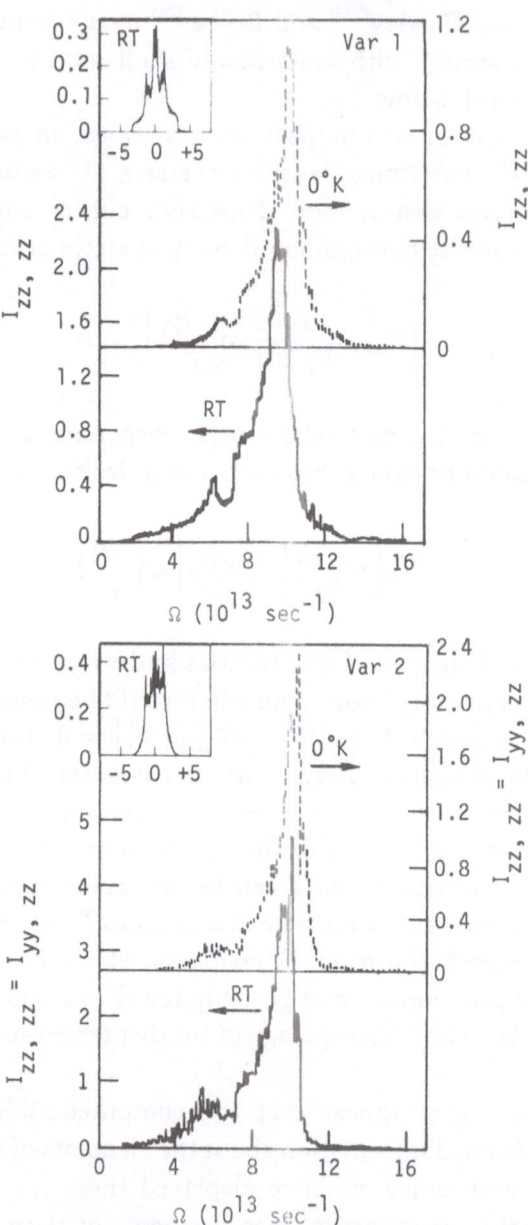

FIG. 44. *See legend opposite.*

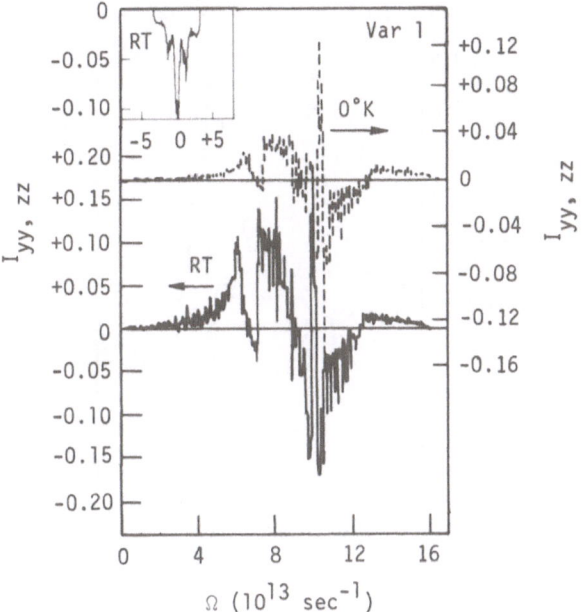

FIG. 44. Curves for NaF of the independent nonzero components $I_{zz,zz}$ and $I_{yy,zz}$ of the second-order Raman intensity computed according to variations 1 and 2 described in the text ($I_{yy,zz} = I_{zz,zz}$ for variation 2). Main curves are the combination bands, including overtones; inset curves are the difference bands, plotted in same units as the main curves. The model used is the DD3N model (deformation dipole with short-range forces extended to include closest-neighbor negative ions).

appropriate to the difference bands differs from that for the overtone and combination bands. Thus it is not too surprising that additional information can be obtained from the difference spectra.

From the foregoing discussion it is evident that the laser Raman spectra of pure alkali halides offer a means of obtaining frequency-spectra and dispersion-curve information that is very much superior to that obtained by examining the infrared-absorption spectra. One would hope that it might be possible by suitable refinement to carry the calculations to the point where the observed Raman spectra can

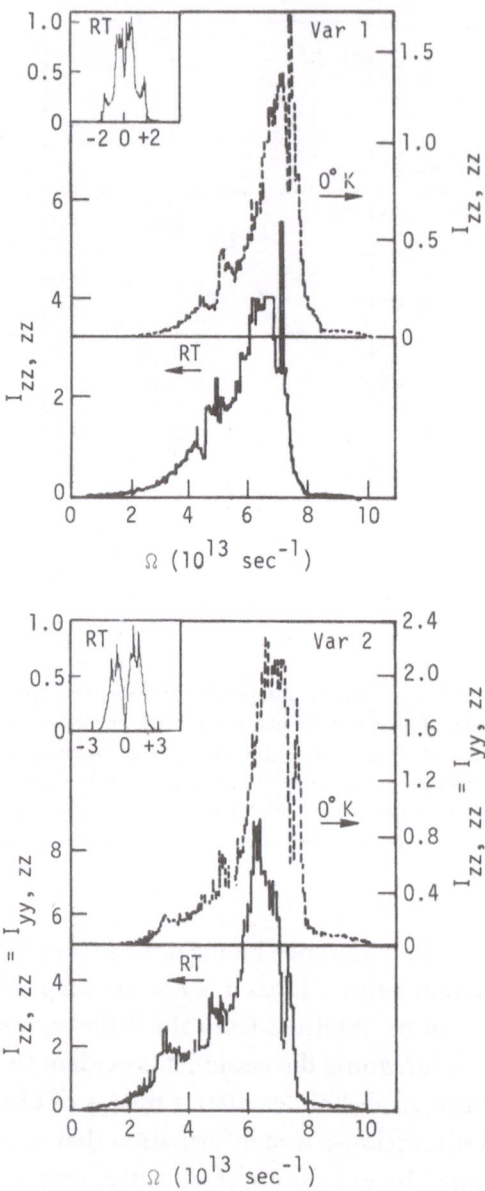

FIG. 45. *See legend opposite.*

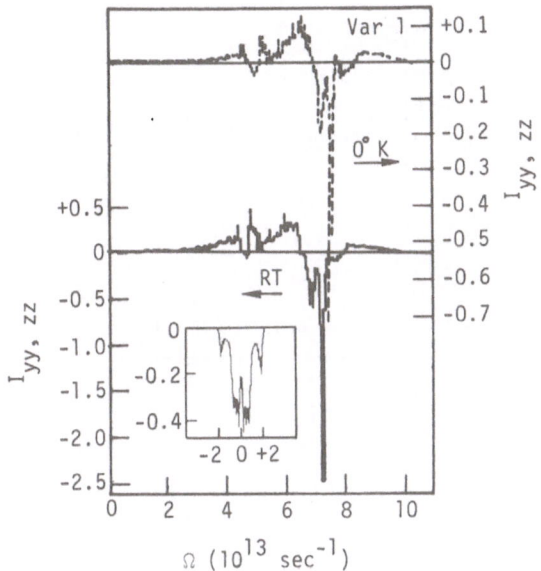

FIG. 45. Curves for NaCl of the independent nonzero components $I_{zz,zz}$ and $I_{yy,zz}$ of the second-order Raman intensity computed according to variations 1 and 2 described in the text ($I_{yy,zz} = I_{zz,zz}$ for variation 2). Main curves are the combination bands, including overtones; inset curves are the difference bands, plotted in same units as the main curves. The model used is the DD3N model (deformation dipole with short-range forces extended to include closest-neighbor negative ions).

be used to study directly the frequency spectra, dispersion curves, and normal-mode eigenvectors of alkali halides—and thereby to use these studies to supplement and refine the information obtained by directly determining phonon-dispersion curves. The success of any such program is contingent on ability to reduce the number of disposable parameters in the polarizabilities by arguments of the type that we have employed for the simple nearest-neighbor model.

At present it would appear that reasonable progress has been made in the interpretation of the second-order Raman spectra.

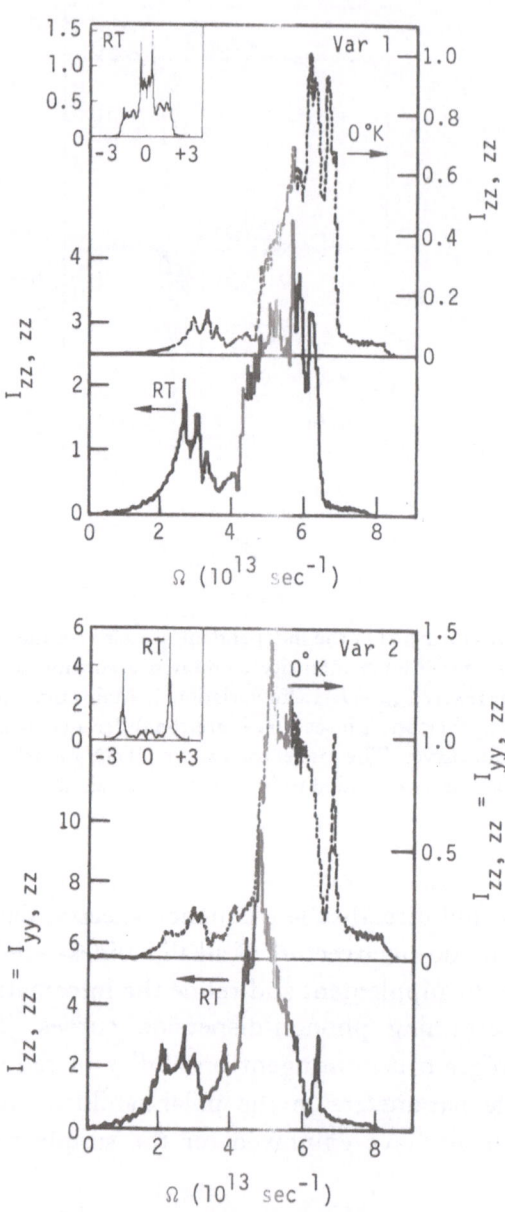

FIG. 46. *See legend opposite.*

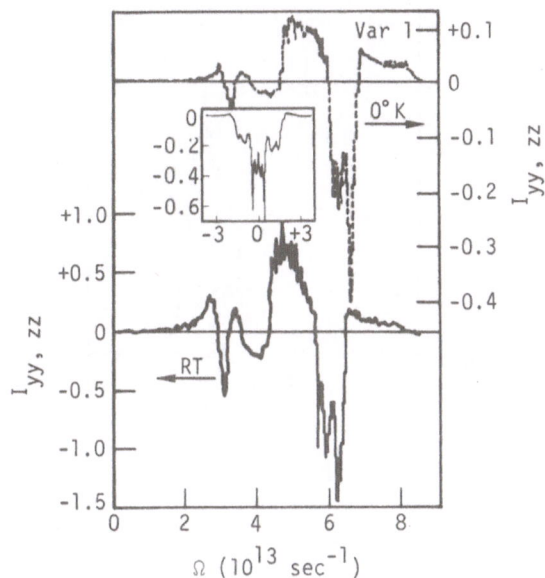

FIG. 46. Curves for NaBr of the independent nonzero components $I_{zz,zz}$ and $I_{yy,zz}$ of the second-order Raman intensity computed according to variations 1 and 2 described in the text ($I_{yy,zz} = I_{zz,zz}$ for variation 2). Main curves are the combination bands, including overtones; inset curves are the difference bands, plotted in same units as the main curves. The model used is the DD3N model (deformation dipole with short-range forces extended to include closest-neighbor negative ions).

However, the theory will remain incomplete until some means can be found of making realistic calculations of the elements of the polarizability tensor from quantum-mechanical perturbation theory.

In this context it would seem essential to develop a more realistic description of the electron–phonon interaction valid for phonons of all wavelengths, since the Fröhlich interaction[28] and the deformation-potential coupling[36] are valid only for very-long-wavelength phonons.

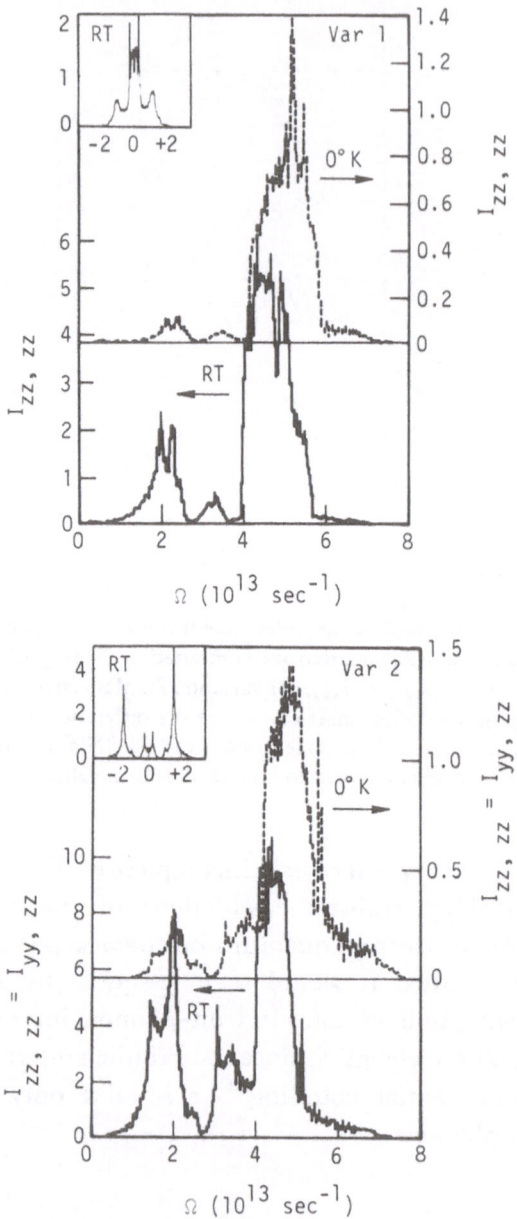

FIG. 47. *See legend opposite.*

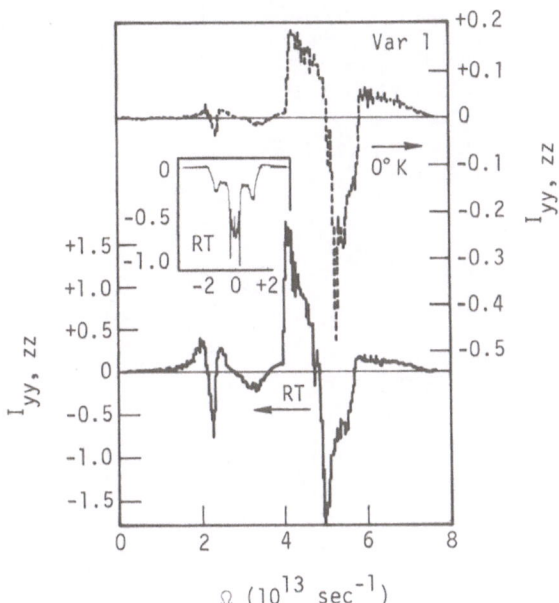

FIG. 47. Curves for NaI of the independent nonzero components $I_{zz,zz}$ and $I_{yy,zz}$ of the second-order Raman intensity computed according to variations 1 and 2 described in the text ($I_{yy,zz} = I_{zz,zz}$ for variation 2). Main curves are the combination bands, including overtones; inset curves are the difference bands, plotted in same units as the main curves. The model used is the DD3N model (deformation dipole with short-range forces extended to include closest-neighbor negative ions).

VI

Dynamic and Static Defects

11. Theory of Impurity Vibrations

Hitherto we have been dealing with the dynamics of perfect alkali halide crystals. We have therefore been able to exploit to the full the lattice periodicity in order to reduce the direct-space force-constant matrix for the whole crystal to a set of independent 6×6 matrices whose eigenvalues give the squares of the normal-mode frequencies. If there is a single impurity present in the lattice, the periodicity is destroyed. It is then necessary, in principle, to return to the full set of direct-space equations of motion and attempt to solve these. In general, this is an intractable problem. However, certain theoretical techniques have been developed in order to circumvent this difficulty. These techniques are based on the fact that a complete solution is not necessary in order to compute experimentally measurable quantities.

The earliest experimental technique used to examine the effects of introducing impurities into insulating crystals was the examination of the associated defect-induced infrared absorption. In this way, one measures a weighted frequency distribution for which the weighting function is some function of the perfect-lattice eigenvectors. More recently there have been extensive studies on defect-activated first-order Raman scattering in alkali halides.

For substitutional defects in alkali halides these two techniques are complementary in that the defect vibrations that are infrared active are not Raman active, and vice versa, within the harmonic approximation. The first observation of lattice-vibrational infrared absorption associated with an impurity appears to have been made by Schäfer,[175] who studied the infrared-absorption spectra due to U centers in a number of alkali halides. These centers are substitutional H^- ions, and since such ions are very much lighter than those of the host lattice, they are capable of vibrations that are highly localized and can, to a good approximation, be regarded as vibrations of the H^- ions alone. These centers have been extensively studied by a number of workers, and it has been found by Fritz[176] and Timusk and Klein[177] that the U-center absorption exhibits side-band structure, which can be associated with anharmonic effects.

The theory of defect vibrations was originally developed in the early 1940s by Lifshitz.[178-180] The final equations obtained in this work are such that very little analytic work can be done with them, unless one assumes a simple Debye spectrum for the frequency distribution of the host lattice. Subsequently Dawber and Elliott[117,181] resurrected and extended this work. Initially they also restricted themselves to a Debye spectrum[117] and proceeded to compute such quantities as the mean-square displacement of the impurity atom. One of the objects of this work was a theoretical prediction of the intensity of the Mössbauer line from ^{57}Fe substituted in various materials. They then went on to consider the infrared absorption associated with a charged isotopic impurity in silicon.[181]

The results of these calculations were compared with experimental results and were found to be in reasonably good agreement. These early calculations were somewhat unrealistic since they treated the impurity as an oscillating point charge. In particular, Leigh and Szigeti[182,183] have questioned the validity of the oscillating-point-charge model and have suggested that a comparable contribution to the absorption arises from the manner in which the point charge polarizes the rest of the lattice, resulting in absorption into modes for which the impurity is at rest. This hypothesis has received experimental support from the work of Newman and Willis[184] on the absorption due to carbon in silicon.

The only isotopic system for which the mass difference of the impurity is large is the LiH–LiD system, which has been studied theoretically by Jaswal and Hardy[135] and by Elliott and Taylor.[185] It has become very clear over the past few years that, for nonisotopic impurities, it is necessary to consider the effect of changed force constants on the defect vibrations.

It is not our intention to present a detailed exposition of the formalism. For our purposes it is sufficient to use the concise exposition of McCombie.[186,187] Our object in doing so is to show that, so long as the only significant force-constant changes occur in the bonds between the defect and its first neighbors, the problem reduces to the computation of certain symmetry-adapted combinations of the perfect-lattice eigenvectors. For illustration we shall present appropriate results obtained with the deformation-dipole model.

The problem we are concerned with is the absorption of energy by the imperfect crystal from an external electromagnetic or other force field. We are considering this problem for the octahedral complex provided by the defect and its six neighbors. We thus have to consider the situation that obtains when generalized driving forces F_t [$F_t \propto \exp(-i\omega t)$] are applied to the complex. Thus, if x_s denotes a symmetry-adapted combination of first-neighbor displacements, we can write

$$x_s = \sum_t [P(\omega) + iQ(\omega)]_{st} F_t \qquad (11.1)$$

where $\mathbf{P} + i\mathbf{Q}$ is the response matrix for the perfect lattice; all relevant information is contained[186,187] in $\mathbf{Q}(\omega)$.

If we express x_s in terms of the normal coordinates $Q\left(\begin{smallmatrix}\mathbf{q}\\j\end{smallmatrix}\right)$, then we have

$$x_s = \sum_{qj} \alpha_s\left(\begin{smallmatrix}\mathbf{q}\\j\end{smallmatrix}\right) Q\left(\begin{smallmatrix}\mathbf{q}\\j\end{smallmatrix}\right)$$

and, for the perfect lattice,

$$x_s = \sum_{qjt} \alpha_s\left(\begin{smallmatrix}\mathbf{q}\\j\end{smallmatrix}\right)\left[\omega^2\left(\begin{smallmatrix}\mathbf{q}\\j\end{smallmatrix}\right) - \omega^2 - i\varepsilon\omega\right]^{-1} \alpha_t\left(\begin{smallmatrix}\mathbf{q}\\j\end{smallmatrix}\right) F_t \qquad (11.2)$$

where we have included an infinitesimal damping to produce dissipation. Now we can regard the effects of mass and force-constant changes as additional "driving forces" on the right-hand side of Eq. (11.2). Thus, for the imperfect lattice,

$$x_s = \sum_{qjtr} \alpha_s\left(\begin{smallmatrix}\mathbf{q}\\j\end{smallmatrix}\right)\left[\omega^2\left(\begin{smallmatrix}\mathbf{q}\\j\end{smallmatrix}\right) - \omega^2 - i\varepsilon\omega\right]^{-1} \alpha_t\left(\begin{smallmatrix}\mathbf{q}\\j\end{smallmatrix}\right)(F_t + \Delta_{tr}x_r) \qquad (11.3)$$

where Δ_{tr} is the defect matrix element that couples the configurations x_t and x_r. Equation (11.3) constitutes a set of linear inhomogeneous equations for the x terms and can be cast in the form of Eq. (11.1). However, we need proceed no further for our purposes because we do not propose to specify the Δ matrix.

The crucial point, as is apparent from Eqs. (11.1) and (11.3), is that the **P** and **Q** matrices are uniquely determined by the products $\alpha_s\left(\begin{smallmatrix}\mathbf{q}\\j\end{smallmatrix}\right) \times \alpha_t\left(\begin{smallmatrix}\mathbf{q}\\j\end{smallmatrix}\right)$ or, for an infinite crystal, by the weighted densities of states:

$$A_{rs}(\Omega) = \lim_{\substack{N\to\infty\\ \Delta\Omega\to 0}} (3N\Delta\Omega)^{-1} \sum_{qj} \int_{\Omega}^{\Omega+\Delta\Omega} \alpha_r\left(\begin{smallmatrix}\mathbf{q}\\j\end{smallmatrix}\right)\alpha_s\left(\begin{smallmatrix}\mathbf{q}\\j\end{smallmatrix}\right)\delta\left[\Omega' - \omega\left(\begin{smallmatrix}\mathbf{q}\\j\end{smallmatrix}\right)\right] d\Omega'$$

$$(11.4)$$

Thus the actual computation reduces to evaluating various projections of the perfect-lattice normal coordinates on the symmetry-adapted normal coordinates of the O_h group. We allow for force-constant changes between the defect and its six first neighbors and assume that these force-constant changes are the same in all six bonds. We also allow for the possibility of the impurity mass being different from that of the host-lattice atom that it replaces. Finally, we shall consider defects of this kind on either sublattice.

It is possible to classify the normal coordinates of such octahedral complexes in accordance with the irreducible representations of the cubic point group:

$$A_{1g}, \qquad E_g, \qquad T_{1g}, \qquad T_{2g}, \qquad T_{1u}, \qquad T_{2u}$$

Thus we have to compute the following 11 functions for each normal mode $\binom{q}{j}$:

A_{1g} $(r = s = 1)$:

$$\left[\alpha_1\binom{q}{j}\right]^2 = 2\left[\sum_\alpha \sigma_\alpha\left(k\Big|\begin{matrix}q\\j\end{matrix}\right)\sin \pi q_\alpha r_0\right]^2 \tag{11.5a}$$

E_g $(r = s = 2)$:

$$\left[\alpha_2\binom{q}{j}\right]^2 = \left[\sigma_x\left(k\Big|\begin{matrix}q\\j\end{matrix}\right)\sin \pi q_x r_0 - \sigma_y\left(k\Big|\begin{matrix}q\\j\end{matrix}\right)\sin \pi q_y r_0\right]^2$$

$$+ \left[\sigma_y\left(k\Big|\begin{matrix}q\\j\end{matrix}\right)\sin \pi q_y r_0 - \sigma_z\left(k\Big|\begin{matrix}q\\j\end{matrix}\right)\sin \pi q_z r_0\right]^2$$

$$+ \left[\sigma_z\left(k\Big|\begin{matrix}q\\j\end{matrix}\right)\sin \pi q_z r_0 - \sigma_x\left(k\Big|\begin{matrix}q\\j\end{matrix}\right)\sin \pi q_x r_0\right]^2 \tag{11.5b}$$

T_{1g} $(r = s = 3)$:

$$\left[\alpha_3\binom{\mathbf{q}}{j}\right]^2 = \left[\sigma_x\left(k\,\Big|\,\begin{matrix}\mathbf{q}\\j\end{matrix}\right) \sin \pi q_y r_0 - \sigma_y\left(k\,\Big|\,\begin{matrix}\mathbf{q}\\j\end{matrix}\right) \sin \pi q_x r_0\right]^2$$

$$+ \left[\sigma_y\left(k\,\Big|\,\begin{matrix}\mathbf{q}\\j\end{matrix}\right) \sin \pi q_z r_0 - \sigma_z\left(k\,\Big|\,\begin{matrix}\mathbf{q}\\j\end{matrix}\right) \sin \pi q_y r_0\right]^2$$

$$+ \left[\sigma_z\left(k\,\Big|\,\begin{matrix}\mathbf{q}\\j\end{matrix}\right) \sin \pi q_x r_0 - \sigma_x\left(k\,\Big|\,\begin{matrix}\mathbf{q}\\j\end{matrix}\right) \sin \pi q_z r_0\right]^2$$

$$(11.5c)$$

T_{2g} $(r = s = 4)$:

$$\left[\alpha_4\binom{\mathbf{q}}{j}\right]^2 = \left[\sigma_x\left(k\,\Big|\,\begin{matrix}\mathbf{q}\\j\end{matrix}\right) \sin \pi q_y r_0 + \sigma_y\left(k\,\Big|\,\begin{matrix}\mathbf{q}\\j\end{matrix}\right) \sin \pi q_x r_0\right]^2$$

$$+ \left[\sigma_y\left(k\,\Big|\,\begin{matrix}\mathbf{q}\\j\end{matrix}\right) \sin \pi q_z r_0 + \sigma_z\left(k\,\Big|\,\begin{matrix}\mathbf{q}\\j\end{matrix}\right) \sin \pi q_y r_0\right]^2$$

$$+ \left[\sigma_z\left(k\,\Big|\,\begin{matrix}\mathbf{q}\\j\end{matrix}\right) \sin \pi q_x r_0 + \sigma_x\left(k\,\Big|\,\begin{matrix}\mathbf{q}\\j\end{matrix}\right) \sin \pi q_z r_0\right]^2$$

$$(11.5d)$$

$T_{1u}(1)$ $(r = s = 5)$:

$$\left[\alpha_5\binom{\mathbf{q}}{j}\right]^2 = 2\sum_\alpha \left[\sigma_\alpha\left(k\,\Big|\,\begin{matrix}\mathbf{q}\\j\end{matrix}\right) \cos \pi q_\alpha r_0\right]^2 \qquad (11.5e)$$

$T_{1u}(2)$ $(r = s = 6)$:

$$\left[\alpha_6\binom{q}{j}\right]^2 = \left[\sigma_x\left(k\Big|\begin{matrix}q\\j\end{matrix}\right)(\cos \pi q_y r_0 + \cos \pi q_z r_0)\right]^2$$

$$+ \left[\sigma_y\left(k\Big|\begin{matrix}q\\j\end{matrix}\right)(\cos \pi q_z r_0 + \cos \pi q_x r_0)\right]^2$$

$$+ \left[\sigma_z\left(k\Big|\begin{matrix}q\\j\end{matrix}\right)(\cos \pi q_x r_0 + \cos \pi q_y r_0)\right]^2 \qquad (11.5f)$$

T_{2u} $(r = s = 7)$:

$$\left[\alpha_7\binom{q}{j}\right]^2 = \left[\sigma_x\left(k\Big|\begin{matrix}q\\j\end{matrix}\right)(\cos \pi q_y r_0 - \cos \pi q_z r_0)\right]^2$$

$$+ \left[\sigma_y\left(k\Big|\begin{matrix}q\\j\end{matrix}\right)(\cos \pi q_z r_0 - \cos \pi q_x r_0)\right]^2$$

$$+ \left[\sigma_z\left(k\Big|\begin{matrix}q\\j\end{matrix}\right)(\cos \pi q_x r_0 - \cos \pi q_y r_0)\right]^2 \qquad (11.5g)$$

T_{1u} $(r = s = 8)$:

$$\left[\alpha_8\binom{q}{j}\right]^2 = \sum_\alpha \left[\sigma_\alpha\left(k\Big|\begin{matrix}q\\j\end{matrix}\right)\right]^2 \qquad (11.5h)$$

Finally we compute three cross terms that mix different α terms having the same symmetry.

T_{1u} $(r = 5; s = 6)$:

$$\alpha_5\binom{\mathbf{q}}{j}\alpha_6\binom{\mathbf{q}}{j} = 2^{1/2}\left\{\left[\sigma_x\left(k\middle|\begin{matrix}\mathbf{q}\\j\end{matrix}\right)\right]^2\right.$$

$$\times\,[\cos \pi q_x r_0\,(\cos \pi q_y r_0 + \cos \pi q_z r_0)]$$

$$+\left[\sigma_y\left(k\middle|\begin{matrix}\mathbf{q}\\j\end{matrix}\right)\right]^2$$

$$\times\,[\cos \pi q_y r_0\,(\cos \pi q_z r_0 + \cos \pi q_x r_0)]$$

$$+\left[\sigma_z\left(k\middle|\begin{matrix}\mathbf{q}\\j\end{matrix}\right)\right]^2$$

$$\left.\times\,[\cos \pi q_z r_0\,(\cos \pi q_x r_0 + \cos \pi q_y r_0)]\right\} \quad (11.5\text{i})$$

T_{1u} $(r = 5; s = 8)$:

$$\alpha_5\binom{\mathbf{q}}{j}\alpha_8\binom{\mathbf{q}}{j} = 2^{1/2}\sum_\alpha \sigma_\alpha\left(k\middle|\begin{matrix}\mathbf{q}\\j\end{matrix}\right)\sigma_\alpha\left(k'\middle|\begin{matrix}\mathbf{q}\\j\end{matrix}\right)\cos \pi q_\alpha r_0$$

$$(k \neq k') \quad (11.5\text{j})$$

T_{1u} $(r = 6; s = 8)$:

$$\alpha_6\binom{\mathbf{q}}{j}\alpha_8\binom{\mathbf{q}}{j} = \left[\sigma_x\left(k\middle|\begin{matrix}\mathbf{q}\\j\end{matrix}\right)\sigma_x\left(k'\middle|\begin{matrix}\mathbf{q}\\j\end{matrix}\right)\right](\cos \pi q_y r_0 + \cos \pi q_z r_0)$$

$$+\left[\sigma_y\left(k\middle|\begin{matrix}\mathbf{q}\\j\end{matrix}\right)\sigma_y\left(k'\middle|\begin{matrix}\mathbf{q}\\j\end{matrix}\right)\right]$$

$$\times\,(\cos \pi q_z r_0 + \cos \pi q_x r_0)$$

(equation continues on next page)

$$(11.5\text{k})$$

(equation (11.5k) continued)

$$+ \left[\sigma_z\!\left(k \bigg| \begin{matrix} \mathbf{q} \\ j \end{matrix} \right) \sigma_z\!\left(k' \bigg| \begin{matrix} \mathbf{q} \\ j \end{matrix} \right) \right]$$

$$\times (\cos \pi q_x r_0 + \cos \pi q_y r_0) \qquad (k \neq k')$$

Thus we have computed the weighted densities of states, $A_{rs}(\Omega)$, for a variety of the rock-salt-structure alkali halide crystals and for various models. A representative result is shown in Table IX for the A_{1g} projection for a positive-ion defect in sodium chloride, while Figs. 48–58 show the computed weighted densities for potassium iodide, potassium chloride, and rubidium chloride. Given such data, one can calculate the response function for any given set of force-constant and mass changes by solving Eq. (11.3).

For ionic crystals, the strength of the infrared absorption should be determined by computing the projection of the particular defect vibration on the $\mathbf{q} = 0$ transverse optic mode. This also can be done by using the appropriate weighted density of states.

For defect-activated Raman scattering, some estimate must be made of the elements of the Raman tensor. This presents a problem similar to that discussed earlier with regard to Raman scattering by the perfect lattice. Some progress has been made,‡ but the shape of the Raman spectrum is of more interest than its intensity, which is an extremely difficult quantity to measure. Raman-active modes have even parity and the defect itself does not vibrate; any Raman activity is entirely due to the electronic mismatch between the defect and the host lattice. Consequently, there are no Raman-active modes for a simple isotopic impurity.

The experimental observation of impurity-induced single-phonon Raman scattering in the alkali halides has many aspects, and there is now available a wealth of data.§

‡ See, for example, Reference 188.
§ See, for example, Reference 189.

TABLE IX. Tabular Data for A_{11}: The Projection of the Normal Modes of the Perfect Crystal on the A_{1g} Vibrations of the Six Nearest Neighbors of a Positive Defect in NaCl[a]

Ω/Ω_{max}	0.001	0.002	0.003	0.004	0.005	0.006	0.007	0.008	0.009	0.010
0.000	0	0	0	0	0	0	0	0	0	0
0.010	0	0	0	0	0	0	0	0	0	0
0.020	0	0	0	0	0	0	0	0	0	0
0.030	0	0	0	0	0	0	0	0	0	0
0.040	0	0	0	0	40	0	0	0	0	0
0.050	6	0	0	0	0	0	40	0	0	0
0.060	0	0	0	1	0	0	0	0	0	0
0.070	9	0	0	159	0	9	0	0	12	0
0.080	0	0	0	2	0	0	0	0	590	3
0.090	0	0	0	17	0	30	0	0	0	0
0.100	40	0	0	0	0	15	2	0	5	0
0.110	0	157	755	0	0	4	0	94	20	782
0.120	0	2	41	41	17	0	0	989	0	4
0.130	1	360	0	23	0	0	0	0	0	164
0.140	30	1 057	1	169	6	637	54	0	65	85
0.150	2	0	1 442	0	2 745	60	0	0	9	2
0.160	105	149	3	196	206	350	33	1 711	72	70
0.170	1 715	137	2	0	652	1 767	0	120	1	0
0.180	1	130	2 228	360	0	345	2 151	21	0	215
0.190	170	309	4 460	2 535	73	1 969	0	205	191	15
0.200	165	60	0	5 219	8	808	0	9	3 134	13
0.210	362	587	2 648	271	0	1 819	1 018	0	426	276
0.220	3 953	5 918	869	2 617	604	509	284	2 762	689	682
0.230	6 624	710	469	183	6 664	907	1	468	7 209	399

TABLE IX (contd.)

Ω/Ω_{max}	0.001	0.002	0.003	0.004	0.005	0.006	0.007	0.008	0.009	0.010
0.240	1 194	827	5	645	17	532	1 377	1 065	7 173	3 800
0.250	5 440	362	0	4 019	9 216	19	448	10 230	7 339	25
0.260	4 818	1 232	212	1 413	1 189	144	1 500	10 158	568	9 740
0.270	10	1 872	2 458	10 652	1 855	6 092	6 217	518	4 185	5
0.280	4 150	1 172	4 478	1 509	6 729	6 846	11 263	16 070	3 422	7 609
0.290	1 968	1 502	15	3 140	16 479	6 305	147	13 590	1 828	8 210
0.300	12 324	5 697	2 421	857	4 993	13 456	4 628	188	19 114	3 304
0.310	18 981	3 583	4 595	4 723	20 042	9 363	5 911	7 622	3 079	15 325
0.320	9 400	30 186	3 084	5 447	3 587	10 436	6 161	4 587	17 643	20 544
0.330	29 376	18 499	8 520	12 678	16 874	11 927	24 926	2 805	5 037	6 826
0.340	8 703	16 109	13 108	15 972	13 958	15 035	50 606	9 138	17 286	25 441
0.350	12 744	4 350	4 253	33 349	24 403	20 002	41 791	19 945	20 016	19 519
0.360	10 675	20 321	36 558	18 289	14 259	17 321	18 931	37 732	29 884	27 260
0.370	23 677	4 748	31 265	35 302	21 265	21 243	5 950	43 335	27 937	41 114
0.380	13 548	60 719	9 929	60 919	36 727	9 675	37 604	47 329	28 718	42 389
0.390	9 775	5 062	78 016	32 565	31 521	17 950	28 676	42 381	33 901	44 038
0.400	43 369	26 158	55 607	59 923	48 749	55 322	34 699	42 354	52 959	47 726
0.410	31 507	39 065	63 349	54 688	46 924	52 167	58 324	45 001	49 852	63 378
0.420	60 433	63 917	93 681	55 936	75 511	43 482	20 472	101 062	101 259	37 018
0.430	40 762	48 780	130 495	69 928	43 803	100 646	109 922	71 672	95 350	73 602
0.440	86 434	70 739	72 440	142 157	90 679	135 368	78 429	98 440	83 500	118 592
0.450	164 894	69 726	144 383	123 967	182 971	133 713	109 120	184 779	105 949	107 703
0.460	168 230	144 115	190 525	198 399	130 048	220 889	142 913	208 712	116 709	256 590
0.470	175 597	177 796	232 859	211 710	172 749	235 176	242 452	343 587	115 309	169 296

TABLE IX *(contd.)*

Ω/Ω_{max}	0.001	0.002	0.003	0.004	0.005	0.006	0.007	0.008	0.009	0.010
0.480	51 630	535 504	120 313	218 719	432 664	423 245	381 607	252 120	318 393	496 410
0.490	510 460	408 353	502 627	371 132	428 267	445 689	382 105	267 013	290 046	460 764
0.500	423 250	598 646	531 292	457 114	229 188	363 425	510 527	365 758	568 329	305 367
0.510	442 851	423 358	747 900	380 613	371 886	469 696	462 200	437 909	545 828	562 490
0.520	490 489	999 999	892 896	444 765	612 807	589 309	511 000	683 161	626 778	499 300
0.530	515 661	598 166	482 435	673 586	678 349	608 314	647 221	483 037	508 083	451 307
0.540	780 310	776 619	840 973	461 849	499 572	489 254	374 995	562 732	394 217	423 379
0.550	474 894	481 289	629 335	230 814	302 464	672 002	310 309	670 393	492 110	395 887
0.560	342 519	447 608	481 628	143 167	166 448	570 185	89 605	492 258	296 346	491 255
0.570	351 658	170 818	249 524	340 102	336 249	426 426	331 123	145 114	561 339	257 736
0.580	294 576	412 666	355 628	337 133	310 076	484 819	178 877	455 968	493 742	461 365
0.590	523 876	436 585	369 398	280 026	669 857	368 123	556 512	454 904	816 245	966 524
0.600	388 701	579 553	461 417	263 095	259 503	499 519	303 502	556 204	502 598	354 318
0.610	483 742	495 244	558 374	270 614	648 361	103 310	238 162	416 917	249 944	425 171
0.620	324 970	413 454	316 098	420 005	393 033	548 840	290 320	374 331	201 672	227 107
0.630	510 618	343 702	492 502	400 578	171 689	551 043	453 767	561 017	809 798	491 250
0.640	198 348	281 011	606 596	378 327	339 860	429 418	335 758	455 877	419 002	339 664
0.650	221 423	453 689	184 902	407 880	203 976	419 200	196 048	160 257	509 556	232 578
0.660	221 642	163 412	266 790	300 768	294 855	140 950	270 413	181 415	363 812	234 264
0.670	123 832	281 147	161 619	138 728	215 565	300 356	138 303	231 742	236 347	125 296
0.680	185 884	179 086	229 468	240 712	60 596	42 844	260 709	174 582	160 985	247 902
0.690	197 299	74 498	242 199	43 344	139 217	292 950	31 417	136 675	64 124	245 803
0.700	149 052	162 196	207 360	138 869	27 941	142 756	101 449	149 270	138 487	124 706
0.710	87 166	201 228	220 332	90 610	187 206	90 128	124 492	68 388	156 207	145 966

TABLE IX (contd.)

Ω/Ω_{max}	0.001	0.002	0.003	0.004	0.005	0.006	0.007	0.008	0.009	0.010
0.720	89 181	129 892	179 081	204 488	233 224	142 815	112 116	141 366	171 351	158 136
0.730	173 381	175 710	202 387	314 185	319 832	166 488	139 638	141 977	76 132	118 241
0.740	71 387	159 133	116 213	43 638	32 503	68 771	16 019	37 492	41 989	73 660
0.750	52 907	28 050	15	22 595	55 673	17 457	29 927	23 655	28 475	34 092
0.760	17 128	17 307	6 819	50 399	30 699	63 018	84 642	48 057	59 310	74 378
0.770	100 538	75 720	76 789	122 570	77 709	158 459	119 882	69 995	139 073	143 546
0.780	236 642	133 087	105 312	79 712	188 646	116 057	85 614	172 188	97 317	107 310
0.790	130 895	61 202	78 065	97 091	116 004	89 091	64 555	89 390	42 000	72 104
0.800	84 044	69 461	48 838	69 297	27 986	13 778	43 006	26 135	19 385	6 095
0.810	42 812	70 055	29 631	40 621	3 440	32 563	43 865	54 228	26 170	19 430
0.820	51 754	4 791	51 092	70 815	32 428	53 126	36 225	15 433	22 844	18 973
0.830	41 337	32 463	65 127	24 887	78 358	50 719	36 366	28 191	28 882	36 643
0.840	53 124	20 474	35 591	15 098	97 075	31 693	25 536	76 251	28 983	62 140
0.850	21 082	13 616	65 684	41 638	52 566	4 495	83 138	18 171	53 457	28 359
0.860	48 990	65 213	34 328	65 264	32 543	43 759	38 014	63 232	14 720	16 816
0.870	53 090	63 036	29 422	42 489	57 348	75 706	30 320	60 270	10 887	44 480
0.880	68 308	10 391	38 458	58 492	39 667	68 828	29 395	31 362	34 089	58 462
0.890	65 514	24 508	56 491	18 069	13 095	82 785	68 109	49 525	960	19 411
0.900	50 354	75 936	48 600	49 117	33 714	18 913	81 014	6 166	57 203	14 724
0.910	40 332	65 311	71 067	22 050	28 579	38 550	26 267	75 873	52 737	29 838
0.920	18 570	39 899	56 913	27 467	30 286	47 081	25 801	31 245	33 430	36 588
0.930	45 281	55 310	38 508	10 364	11 025	19 812	72 449	41 397	16 565	25 576
0.940	23 193	5 798	33 031	63 840	16 263	22 137	19 648	19 172	13 559	22 712

TABLE IX (*contd.*)

Ω/Ω_{max}	0.001	0.002	0.003	0.004	0.005	0.006	0.007	0.008	0.009	0.010
0.950	42 912	19 860	41 155	0	11 172	23 509	13 122	37 715	9 013	27 930
0.960	399	18 440	21 532	30 806	3 709	2 363	10 837	24 499	21 001	5 345
0.970	3 325	12 435	14 399	12 453	6 431	4 627	9 512	14 176	5 824	2 275
0.980	3 853	11 374	7 734	875	4 170	5 265	5 744	685	5 080	3 350
0.990	2 694	1 647	1 441	2 577	0	1 505	394	763	252	230

[a] Deformation-dipole model including next-nearest-neighbor short-range forces; low-temperature input data; $\Omega_{max} = 4.9903\,2883 \times 10^{13}$ sec^{-1}; channel coefficient $= 5.8909\,7425 \times 10^{-9}$.

Another possible application of these weighted densities of states is to the computation of impurity effects on lattice thermal conductivity. Calculations of this sort have been made by McCombie and Slater,[190] Elliott and Taylor,[191] and other workers.[192] It is extremely important to allow for the modification of the perfect-lattice modes by the presence of the impurity, and it is not adequate to treat the impurity simply as a scattering center whose cross section can be calculated by using the first Born approximation.[193]

Unfortunately, comparison between theory and experiment is not particularly easy, because it is necessary to separate from the thermal resistivity the part that is actually due to impurity scattering. This is difficult because of the problems associated with making a good estimate of the intrinsic lattice thermal resistivity.

In general, it is possible to obtain semiquantitative agreement between theory and experiment. Thus, when there is a resonance in the defect amplitude somewhere in the range of frequencies spanned by the vibrational spectrum of the host lattice, there is an associated resonance in the phonon-scattering cross section, and this resonance manifests itself as a dip in the plot of thermal conductivity versus temperature. However, there is no simple relationship between the temperature at which this dip occurs and the frequency of the resonance.[192] In spite of this, lattice thermal conductivity has the virtue of being influenced by all types of resonant scattering and thus can be used as a broad-band probe to determine the presence of any type of resonance.

It is not our intention to give in this section an exhaustive account of defect vibrations, since this has been extensively covered by a number of reviews.‡ What we have done in this section is to present the results of realistic calculations of the weighted densities of states, from which it is possible to calculate the various physical effects of impurity mismatch. Hence one can compute the shapes of defect-activated infrared-absorption and Raman-scattering spectra.

‡ See, for example, Reference 194.

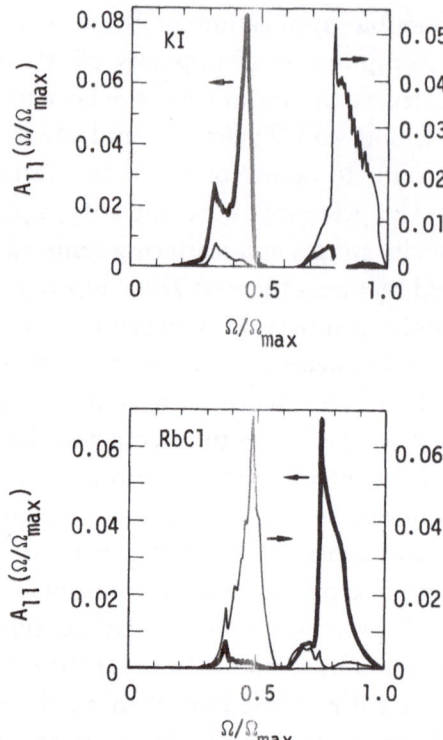

FIG. 48. Plots of the weighted densities of states $A_{11}(\Omega)$ defined by Eq. (11.4), using the projections $\alpha_r(^q_k)$ and $\alpha_s(^q_k)$ defined by Eqs. (11.5a)–(11.5k). Heavy curves ($k = 2$) denote results appropriate to the positive ion located at the origin; light curves ($k = 1$) denote results appropriate to the negative ion located at the origin. The deformation-dipole model with first neighbor repulsive forces has been used,

The manner in which we have presented our results (see Table IX, Figures 48–58, and Ref. 195) is designed to cope with the problem of presenting a vast amount of data in tractable form. The nature of our calculations is such that they generate more data than can be communicated to, or used by, other workers in any simple manner. For example, the calculations of the weighted densities of states just presented were made for samples of 64,000 wave vectors in

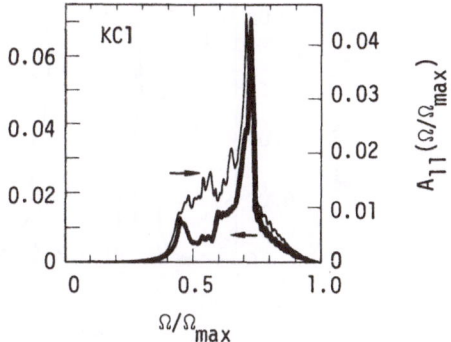

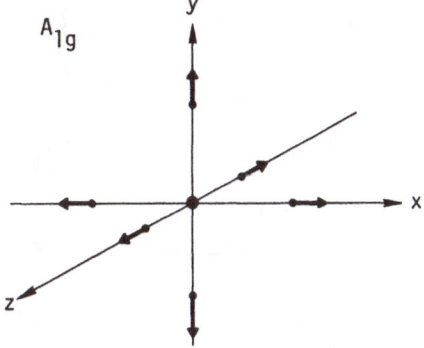

and the calculations have been carried out for low temperature input data. However, the transverse optical frequency has been fitted and the ionic charge reduced to optimize the fit to the measured dispersion curves ($0.90e$ for KI, $0.93e$ for KCl, and $0.95e$ for RbCl).

the reduced zone; these in turn were generated from a basic irreducible set of 1686 wave vectors. If one were to attempt to present the eigenfrequencies and eigenvectors for every member of the irreducible set, the amount of data so generated would be prohibitive. Thus it would be simpler for someone who wished to use these data to reprogram the basic dynamical matrix and generate the eigenvectors and eigenfrequencies. However, this type of computational

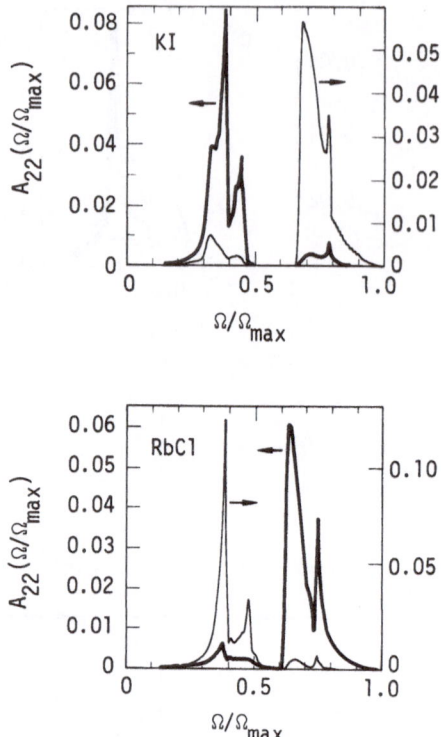

FIG. 49. Plots of the weighted densities of states $A_{22}(\Omega)$ defined by Eq. (11.4), using the projections $\alpha_r\binom{q}{j}$ and $\alpha_s\binom{q}{j}$ defined by Eqs. (11.5a)–(11.5k). Heavy curves ($k = 2$) denote results appropriate to the positive ion located at the origin; light curves ($k = 1$) denote results appropriate to the negative ion located at the origin. The deformation-dipole model with first-neighbor repulsive forces has been used, and

facility is not necessarily available to all workers in this field, and we have therefore attempted to reduce the vast amount of numerical output to tractable proportions. In particular, the numerical data in Table IX could be repunched for use in some other program to be run on a relatively small machine.

We have considered only the effects of changes in nearest-neighbor force constants, although in the future it may be desirable

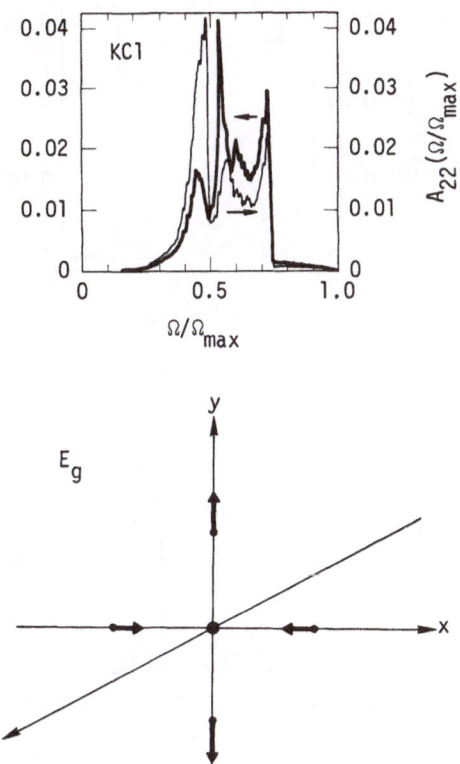

the calculations have been carried out for low-temperature input data. However, the transverse optical frequency has been fitted and the ionic charge reduced to optimize the fit to the measured dispersion curves ($0.90e$ for KI, $0.93e$ for KCl, and $0.95e$ for RbCl).

to consider the corresponding effects for more distant neighbors. The main purpose of the work described in this section is to provide a means of making calculations in which force-constant changes between at least first neighbors are explicitly included. To our knowledge, the first work of this kind was done by McCombie *et al.*,[196] and it is now clearly established that the effects of force-constant changes are often much more important than the effects of changed mass.

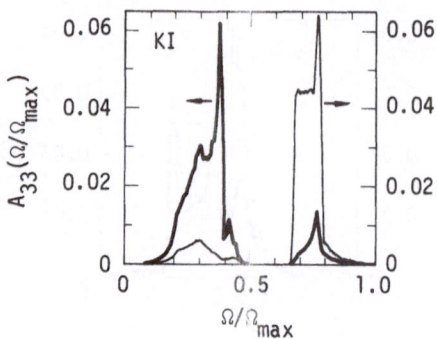

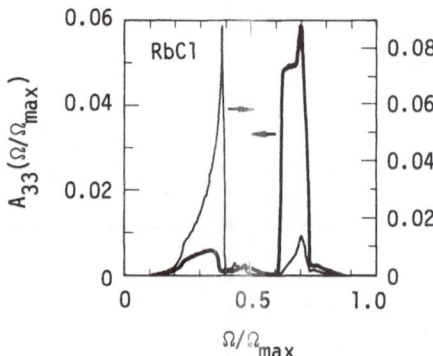

FIG. 50. Plots of the weighted densities of states $A_{33}(\Omega)$ defined by Eq. (11.4), using the projections $\alpha_r(^q_q)$ and $\alpha_t(^q_q)$ defined by Eqs. (11.5a)–(11.5k). Heavy curves ($k = 2$) denote results appropriate to the positive ion located at the origin; light curves ($k = 1$) denote results appropriate to the negative ion located at the origin. The deformation-dipole model with first-neighbor repulsive forces has been used, and

12. The Method of Lattice Statics

We now wish to turn from the dynamical problems we have been discussing to the consideration of static problems. At first sight it might appear that there is no relationship between the two, but it is our basic objective in this section to demonstrate that this is not the case. We shall show that the techniques employed in the study of

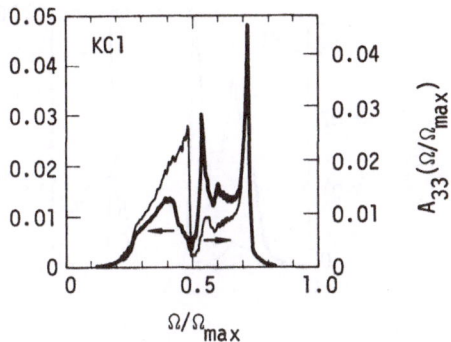

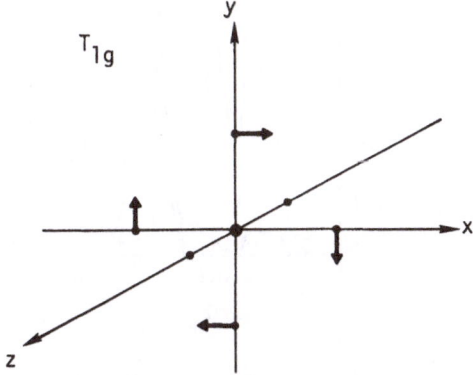

the calculations have been carried out for low-temperature input data. However, the transverse optical frequency has been fitted and the ionic charge reduced to optimize the fit to the measured dispersion curves (0.90e for KI, 0.93e for KCl, and 0.95e for RbCl).

dynamical problems, if suitably modified, can be used in treating the static problems that arise in connection with the study of lattice imperfections. In the preceding section we have already discussed the influence of point imperfections on crystal dynamics, but this is only one aspect of defect problems and not necessarily the most important.

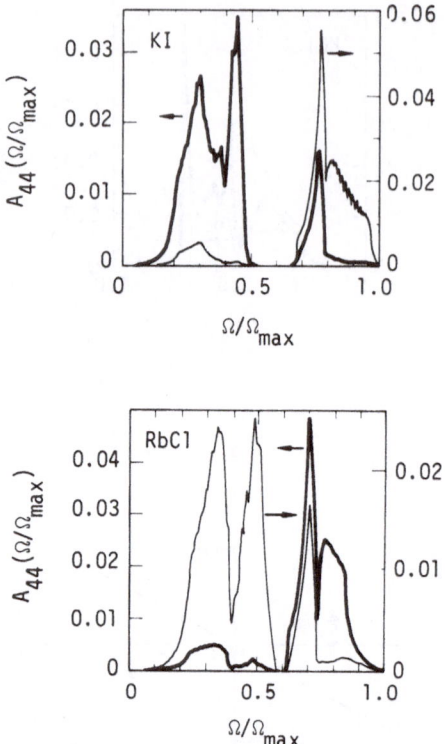

FIG. 51. Plots of the weighted densities of states $A_{44}(\Omega)$ defined by Eq. (11.4), using the projections $\alpha_r(^q_j)$ and $\alpha_s(^q_j)$ defined by Eqs. (11.5a)–(11.5k). Heavy curves ($k = 2$) denote results appropriate to the positive ion located at the origin; light curves ($k = 1$) denote results appropriate to the negative ion located at the origin. The deformation-dipole model with first-neighbor repulsive forces has been used, and

An imperfection in a crystal lattice (we confine ourselves to point imperfections) causes the lattice to distort and the ions to polarize. This distortion and polarization are, at least in principle, measurable by such techniques as nuclear quadrupole resonance, which probes the field gradients at ion sites in the vicinity of an impurity.[197] These field gradients are determined by the distortions and polarizations of the ions in the lattice.

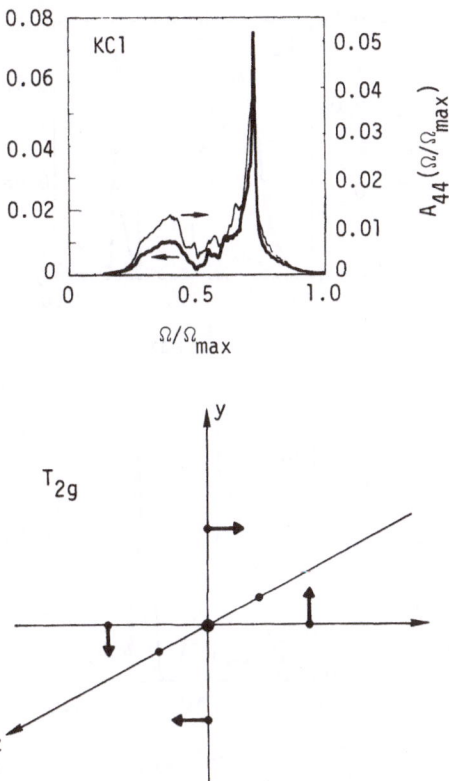

the calculations have been carried out for low-temperature input data. However, the transverse optical frequency has been fitted and the ionic charge reduced to optimize the fit to the measured dispersion curves (0.90e for KI, 0.93e for KCl, and 0.95e for RbCl).

The field gradient is directly measurable, but the indirect influence of these distortions and polarizations is often much more important. Specifically, it is of critical importance from many points of view (e.g., radiation-damage studies, color-center studies, interpretation of ionic conductivity) to know the formation energy of the defect. This is the energy required to create the defect and is normally defined, at least for lattice vacancies, as the energy required

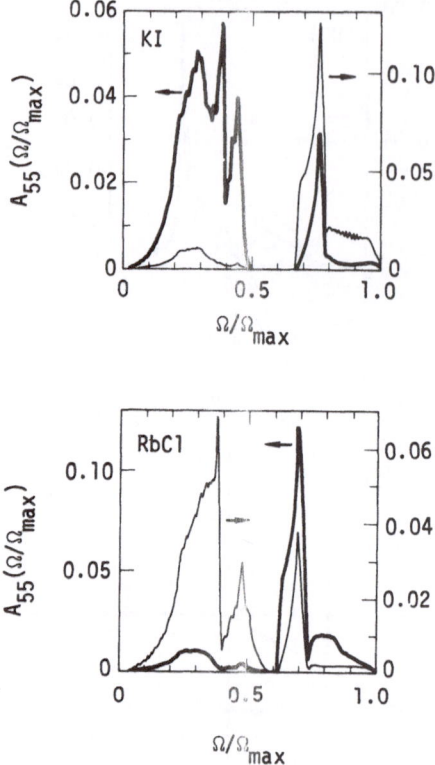

FIG. 52. Plots of the weighted densities of states $A_{55}(\Omega)$ defined by Eq. (11.4), using the projections $\alpha_r\binom{q}{i}$ and $\alpha_s\binom{q}{i}$ defined by Eqs. (11.5a)–(11.5k). Heavy curves ($k = 2$) denote results appropriate to the positive ion located at the origin; light curves ($k = 1$) denote results appropriate to the negative ion located at the origin. The deformation-dipole model with first-neighbor repulsive forces has been used, and

to remove an ion from an interior site and place it on the surface. We shall restrict our attention to vacancies. For these defects the calculated formation energy (if the lattice is held rigid and the ions not allowed to polarize) is larger than the observed value by a factor on the order of 3–4. The reason for this discrepancy is that in reality the lattice both relaxes and polarizes when the defect is present. Both of

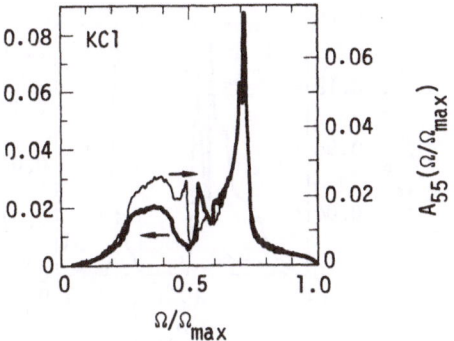

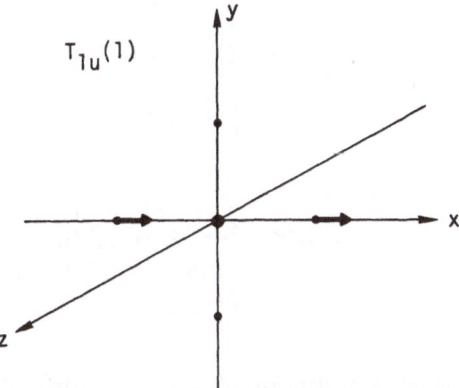

the calculations have been carried out for low-temperature input data. However, the transverse optical frequency has been fitted and the ionic charge reduced to optimize the fit to the measured dispersion curves ($0.90e$ for KI, $0.93e$ for KCl, and $0.95e$ for RbCl).

these types of relaxation contribute to reducing the theoretical formation energy. Since this reduction is evidently very large, it is of critical importance to be able to calculate this relaxation energy with the same precision as one can calculate the static lattice energy, which determines the formation energy for the rigid lattice.

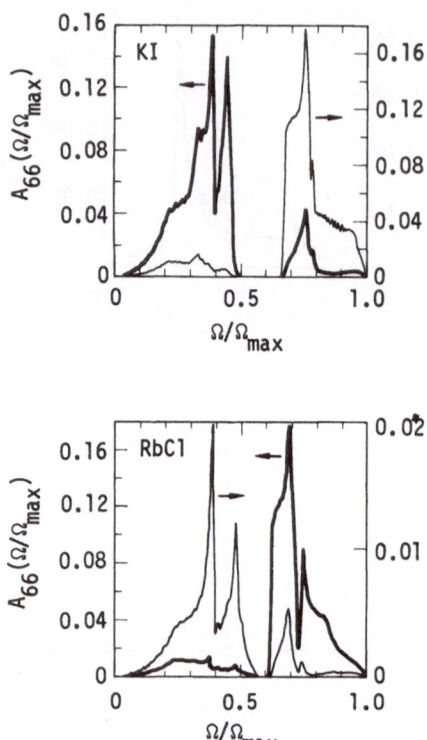

FIG. 53. Plots of the weighted densities of states $A_{66}(\Omega)$ defined by Eq. (11.4), using the projections $\alpha_r({}^q_j)$ and $\alpha_t({}^q_j)$ defined by Eqs. (11.5a)–(11.5k). Heavy curves ($k = 2$) denote results appropriate to the positive ion located at the origin; light curves ($k = 1$) denote results appropriate to the negative ion located at the origin. The deformation-dipole model with first-neighbor repulsive forces has been used, and

In the past, the standard method of calculating this relaxation energy for charged defects in an ionic crystal has been that originally developed by Mott and Littleton in 1938.[198] This is a particular case of what we shall refer to as semidiscrete methods, which are used to treat many types of defect problems. The basis of these methods is to treat the defect and a finite number of its close neighbors—which we shall refer to as belonging to region I—as discrete, that is, on an

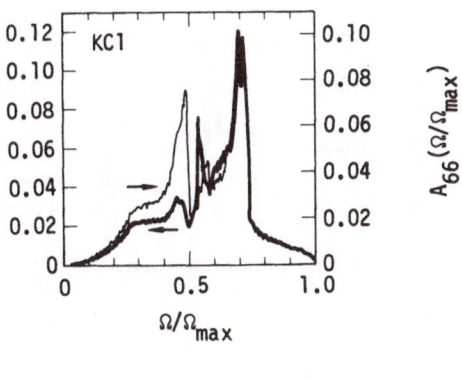

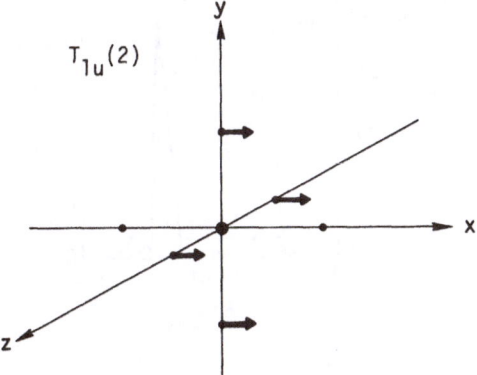

the calculations have been carried out for low-temperature input data. However, the transverse optical frequency has been fitted and the ionic charge reduced to optimize the fit to the measured dispersion curves (0.90e for KI, 0.93e for KCl, and 0.95e for RbCl).

atomistic basis. Outside this region lies the remainder of the lattice, which we call region II and which is treated as an elastic and dielectric continuum. Another example of this type of approach is provided by Johnson's[199] calculations on defects in body-centered-cubic metals. In his work, region II is solely an elastic continuum.

This type of approach inevitably runs into the problem of matching the continuum displacement field in region II to the

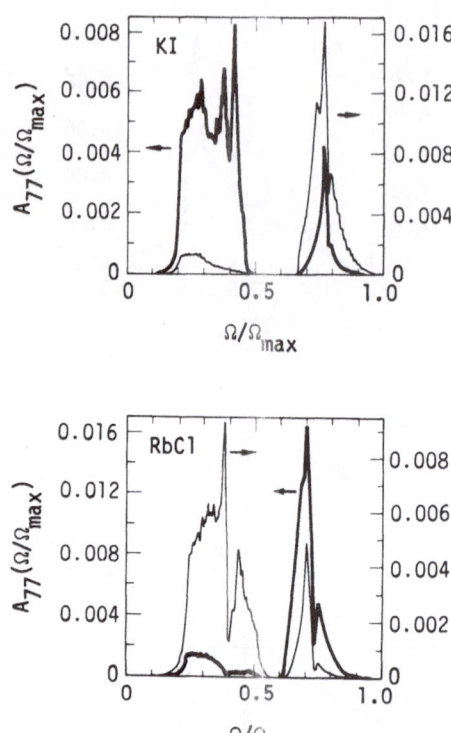

FIG. 54. Plots of the weighted densities of states $A_{77}(\Omega)$ defined by Eq. (11.4), using the projections $\alpha_r\binom{q}{i}$ and $\alpha_t\binom{q}{i}$ defined by Eqs. (11.5a)–(11.5k). Heavy curves ($k = 2$) denote results appropriate to the positive ion located at the origin; light curves ($k = 1$) denote results appropriate to the negative ion located at the origin. The deformation-dipole model with first-neighbor repulsive forces has been used, and

displacements calculated for the outer atoms of region I. It may be that the defect-formation energy is not too sensitive to the details of the matching. However, the calculations made by the approach we shall describe indicate that the mismatch at the interface between regions I and II is usually very marked.

The Mott–Littleton technique provides the simplest example of the semidiscrete method. In the original calculation, the only ions

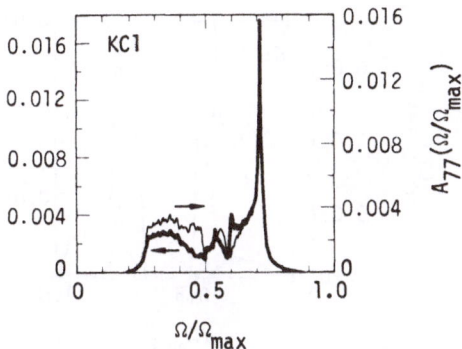

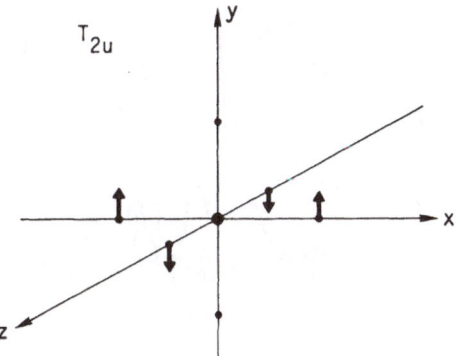

the calculations have been carried out for low-temperature input data. However, the transverse optical frequency has been fitted and the ionic charge reduced to optimize the fit to the measured dispersion curves ($0.90e$ for KI, $0.93e$ for KCl, and $0.95e$ for RbCl).

included in region I are the six first neighbors of the defect. The remainder of the lattice is treated as a polarizable dielectric continuum, and the total ionic dipole moments of region II are computed by dividing the macroscopic electrical polarization

$$\mathbf{P} = (4\pi)^{-1}(1 - \varepsilon_0^{-1})(z e \mathbf{r}/r^3) \tag{12.1}$$

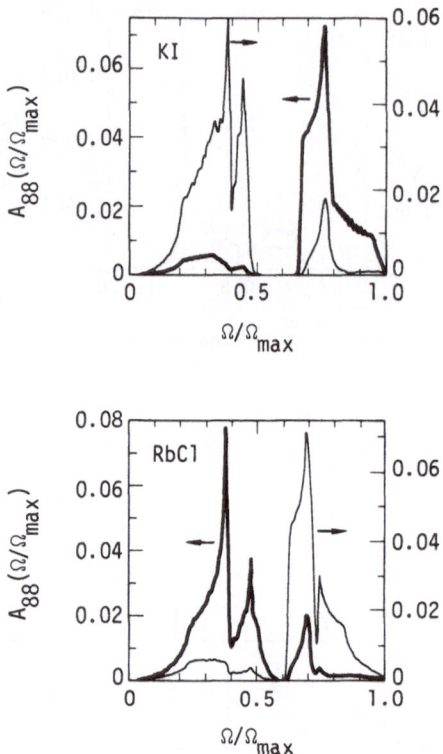

FIG. 55. Plots of the weighted densities of states $A_{88}(\Omega)$ defined by Eq. (11.4), using the projections $\alpha_r\binom{q}{j}$ and $\alpha_s\binom{q}{j}$ defined by Eqs. (11.5a)–(11.5k). Heavy curves ($k = 1$) denote results appropriate to the positive ion located at the origin; light curves ($k = 2$) denote results appropriate to the negative ion located at the origin. The deformation-dipole model with first-neighbor repulsive forces has been used, and

in proportion to the electronic and displacement polarizabilities of the ions. In Eq. (12.1), ze is the effective charge on the defect and r is the distance from the defect; the displacement polarizability is calculated by finding the relative movement of anions and cations in a uniform impressed electric field.

Mott and Littleton[198] justified this approach on the basis of a calculation made for a model lattice in which the ions were allowed to

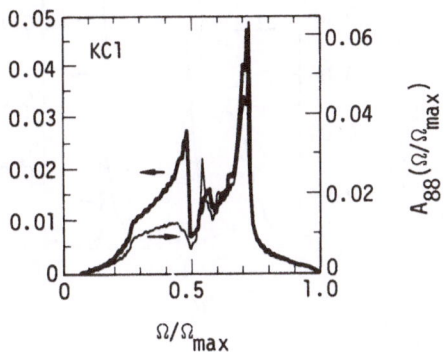

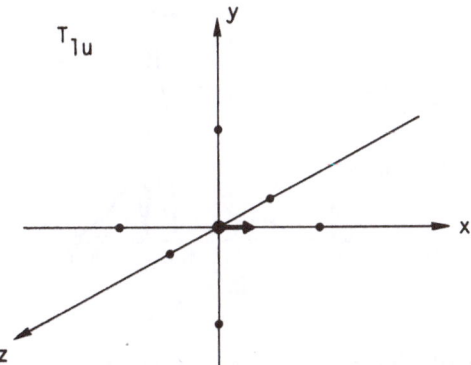

the calculations have been carried out for low-temperature input data. However, the transverse optical frequency has been fitted and the ionic charge reduced to optimize the fit to the measured dispersion curves ($0.90e$ for KI, $0.93e$ for KCl, and $0.95e$ for RbCl). Note also the interchange of the indices $k = 1$ and $k = 2$.

polarize, but not to displace. As they treated successive shells discretely, they found that in their first approximation—in which they treated the first-neighbor shell as discrete and regarded the rest of the lattice as a continuum—there was a marked change in the value of the formation energy compared with that obtained by treating the whole of the lattice as a continuum. However, inclusion of further shells of neighbors made very little difference. Subsequently, other

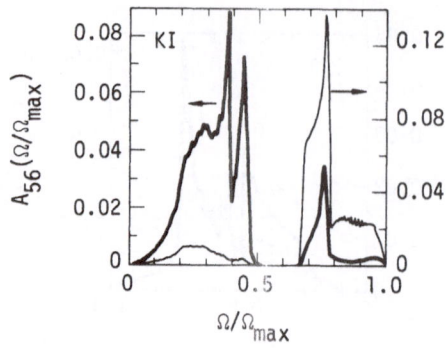

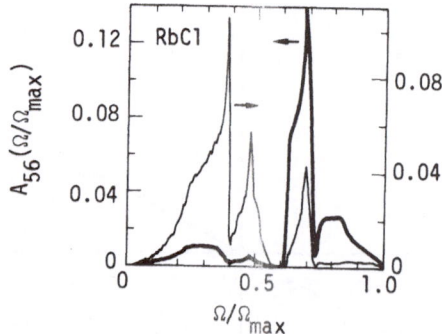

FIG. 56. Plots of the weighted densities of states $A_{56}(\Omega)$ defined by Eq. (11.4), using the projections $\alpha_r(^q_l)$ and $\alpha_t(^q_l)$ defined by Eqs. (11.5a)–(11.5k). Heavy curves ($k = 2$) denote results appropriate to the positive ion located at the origin; light curves ($k = 1$) denote results appropriate to the negative ion located at the origin. The deformation-dipole model with first-neighbor repulsive forces has been used, and

workers extended this calculation as far as the sixth-neighbor shell.[200,201] They confirmed Mott and Littleton's conclusion regarding the formation energy, but they also found that, even for the sixth-neighbor shell, the ionic dipole moments did not agree with those computed by using the continuum approximation.

Recent calculations, based on the Mott–Littleton approach but allowing for a significantly larger region I, have been carried out[202]

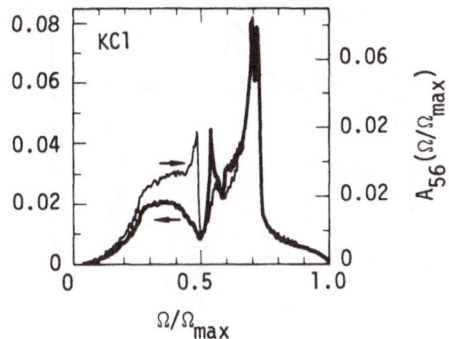

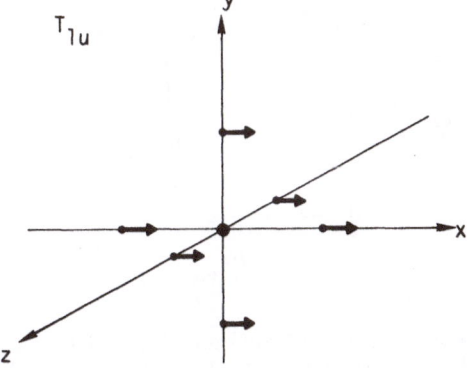

the calculations have been carried out for low-temperature input data. However, the transverse optical frequency has been fitted and the ionic charge reduced to optimize the fit to the measured dispersion curves ($0.90e$ for KI, $0.93e$ for KCl, and $0.95e$ for RbCl).

for vacancies in the whole sequence of alkali halides. Somewhat earlier, Boswarva and Lidiard[203] made a sequence of calculations designed to test the sensitivity of Mott–Littleton results to (1) the form of the non-Coulombic short-range potentials in the lattice and (2) the approximations made in the computation.

In particular, Boswarva and Lidiard studied the effect of using different continuum approximations for region II. The problem that

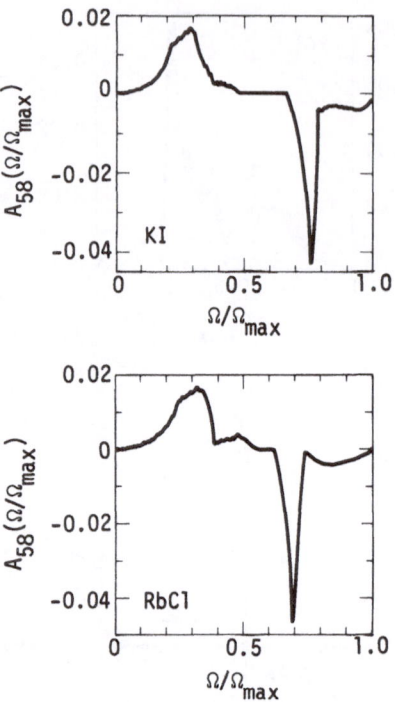

FIG. 57. Plots of the weighted densities of states $A_{58}(\Omega)$ defined by Eq. (11.4), using the projections $\alpha_r\binom{q}{j}$ and $\alpha_s\binom{q}{j}$ defined by Eqs. (11.5a)–(11.5k). In this case the results appropriate to the positive ion located at the origin ($k = 2$) and those appropriate to the negative ion located at the origin ($k = 1$) are identical. The deformation-dipole model with first-neighbor repulsive forces has been used, and

arises in this context is a consequence of the need to allow for elastic displacements in region II in addition to the polarization displacements given by the Mott–Littleton prescription. The necessity of including such a displacement field was first pointed out by Brauer,[204] who suggested matching a continuum solution of the form

$$\mathbf{u} = c(\mathbf{r}/r^3) \tag{12.2}$$

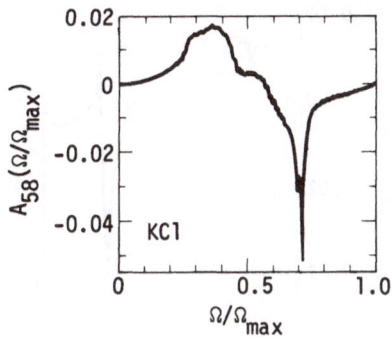

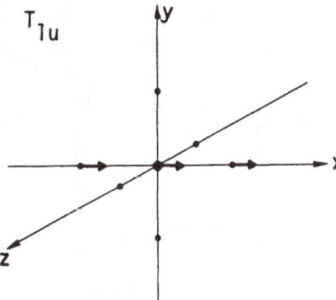

the calculations have been carried out for low-temperature input data. However, the transverse optical frequency has been fitted and the ionic charge reduced to optimize the fit to the measured dispersion curves ($0.90e$ for KI, $0.93e$ for KCl, and $0.95e$ for RbCl).

to the first-neighbor displacements. This is almost certainly an overestimate, and Boswarva and Lidiard also used an intermediate approximation which matched the elastic solution to the part of the first-neighbor displacement they regarded as representing the "elastic component." They found their calculations were sensitive to changes of this nature. Alternative work by Scholz[205] considered only region I, but made it much larger, so that it contained 256 ions.

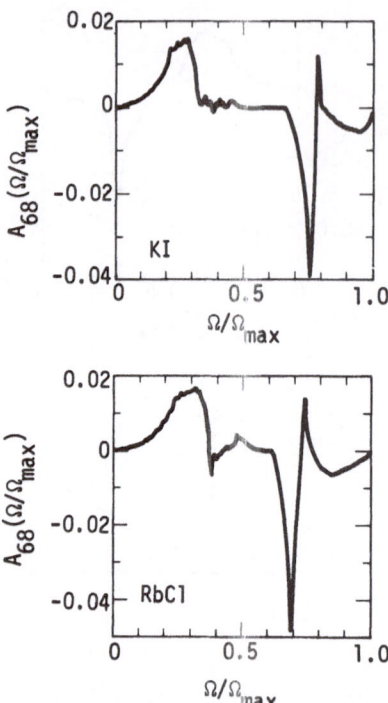

FIG. 58. Plots of the weighted densities of states $A_{68}(\Omega)$ defined by Eq. (11.4), using the projections $\alpha_r\binom{q}{j}$ and $\alpha_t\binom{q}{j}$ defined by Eqs. (11.5a)–(11.5k). In this case the results appropriate to the positive ion located at the origin ($k = 2$) and those appropriate to the negative ion located at the origin ($k = 1$) are identical. The deformation-dipole model with first-neighbor repulsive forces has been used, and

Calculations made by using the extended Mott–Littleton techniques are somewhat cumbersome, and in particular the dichotomy between regions I and II is cause for concern. Thus one would like an approach that is free from this dichotomy and provides a means of determining the asymptotic polarization and displacement fields at large distances from the defect. This is provided by the method of lattice statics, whose name has its origin in the similarity between the

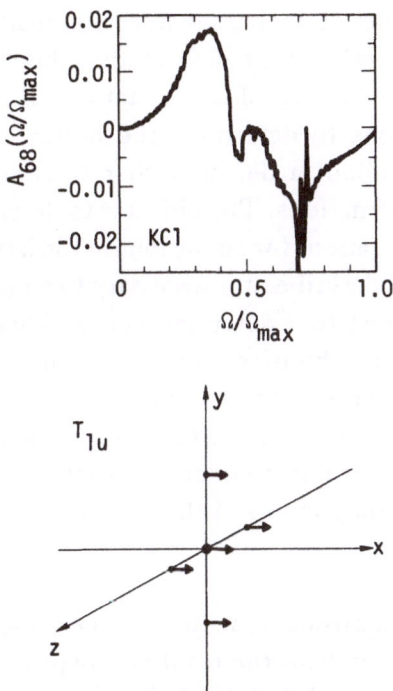

the calculations have been carried out for low-temperature input data. However, the transverse optical frequency has been fitted and the ionic charge reduced to optimize the fit to the measured dispersion curves ($0.90e$ for KI, $0.93e$ for KCl, and $0.95e$ for RbCl).

techniques used and those employed in the theory of lattice dynamics. It was originally formulated by Kanzaki,[206,207] who selected the simplest possible problem to handle by this new technique and considered a vacancy in solid argon. He used a Lennard-Jones 6–12 potential to describe the interatomic potential.

Subsequently the new technique was extended by one of us (J.R.H.)[208,209] to the treatment of a substitutional potassium ion in sodium chloride. These calculations were first made with a rigid-ion model and subsequently with a deformation-dipole model.

The basis of the method of lattice statics is the use of the lattice equilibrium equations to determine the lattice configuration and ionic polarizations about a defect, rather than any continuum or semidiscrete approximations. The object is to determine the explicit solutions of these equations for all the ions in the lattice. To do this in the manner we shall describe, it is necessary to employ the harmonic approximation for the lattice potential energy. When this is done, it is possible to make use of Fourier transformations of force constants, displacement fields, and so on. In this way, the equilibrium equations, which for direct space form a $6N \times 6N$ array, are reduced to N decoupled sets of 6×6 linear equations, each of which determines one of the Fourier components of the displacement or the polarization fields.

The presence of the defect introduces inhomogeneous terms into the equations of lattice dynamics. Since these inhomogeneous terms are time independent, the resultant displacements are purely static. Thus one can say that the defect exerts generalized forces on the normal coordinates of the perfect lattice.

In considering a charged defect in an ionic crystal, it is necessary to consider the equilibrium equations for the dipole moments on the ions as well as those governing their displacements. In this way, we determine both the equilibrium dipole moments and the equilibrium displacements; we then use the equilibrium equations to transform the relaxation energy into its final form.

The formal development of the theory,[96] provided that we write our equations in matrix form, can be carried out using the direct-space equations. The Fourier transformations can then be carried out very simply because the resultant equations have the same form as the direct-space equations. However, the significance of the various matrices and vectors is changed since they now refer to Fourier components of vectors or Fourier transforms of matrices.

In this way it is possible to obtain a very simple expression for what we shall refer to as the zero-order relaxation energy. To this order one evaluates the effective forces that the defect exerts on the ions of the lattice at the unrelaxed positions of the ions.

We shall now proceed to demonstrate how the method is applied. We begin by taking the Hamiltonian that was used previously for the deformation-dipole model, together with the inhomogeneous term that represents the defect force array. Thus the energy function can be written as

$$W = X(\text{I}:\boldsymbol{\xi}, \boldsymbol{\mu}) + Y(\boldsymbol{\xi}, \boldsymbol{\mu}) \tag{12.3}$$

where $\boldsymbol{\xi}$ and $\boldsymbol{\mu}$ denote the displacements and dipole moments in region II. The symbol X represents the self-energy of region I plus interaction terms between regions I and II. The term $Y(\boldsymbol{\xi}, \boldsymbol{\mu})$ is the potential energy of a region II filled with a perfect unpolarized region I; there are no linear terms in $Y(\boldsymbol{\xi}, \boldsymbol{\mu})$. If we now write out X and Y explicitly, the energy function Y is given by [see Eq. (5.15)]

$$Y = \tfrac{1}{2}\tilde{\boldsymbol{\xi}}(\mathbf{R} - \mathbf{H})\boldsymbol{\xi} - \tfrac{1}{2}\tilde{\boldsymbol{\mu}}'\mathbf{UH}\boldsymbol{\xi} - \tfrac{1}{2}\tilde{\boldsymbol{\xi}}\mathbf{HU}\boldsymbol{\mu}' - \tfrac{1}{2}\tilde{\boldsymbol{\mu}}'\mathbf{UHU}\boldsymbol{\mu}' + \tfrac{1}{2}\tilde{\boldsymbol{\mu}}\mathbf{a}^{-1}\boldsymbol{\mu} \tag{12.4}$$

where

$$\boldsymbol{\mu}' = \boldsymbol{\mu} + \boldsymbol{\mu}_d$$

$$= \boldsymbol{\mu} + \check{\mathbf{S}}\boldsymbol{\xi} \tag{12.4a}$$

The equilibrium values of $\boldsymbol{\mu}$ and $\boldsymbol{\xi}$ are then obtained by setting

$$\frac{\partial W}{\partial \mu_\lambda} = 0 = \frac{\partial W}{\partial \xi_\lambda} \tag{12.5}$$

for all λ. Ultimately, one has to write out the resultant concise results in the explicit notation of Born and Huang.

The equilibrium displacements and dipole moments are determined by minimizing the total energy W with respect to both dipole moments and displacements. Thus we obtain

$$[\mathbf{R} - (1 + \mathbf{SU})\mathbf{H}(1 + \mathbf{U\check{S}})]\xi - (1 + \mathbf{SU})\mathbf{HU}\mu = \mathbf{T}_\xi \quad (12.6)$$

and

$$(\mathbf{a}^{-1} - \mathbf{UHU})\mu - \mathbf{UH}(1 + \mathbf{U\check{S}})\xi = \mathbf{T}_\mu \quad (12.7)$$

where $\mathbf{T}_\xi$ and $\mathbf{T}_\mu$ are the column matrices of the forces:

$$T_\xi^\lambda = \frac{-\partial X}{\partial \xi_\lambda}$$
$$T_\mu^\lambda = \frac{-\partial X}{\partial \mu_\lambda} \quad (12.8)$$

The only forces on the region II dipoles are electrical, so that $\mathbf{T}_\mu$ is a matrix of electrical field strengths acting on region II. To emphasize this we shall replace $\mathbf{T}_\mu$ by the symbol $\mathbf{E}$. We are thus left with two sets of equations involving the dipole moments and the displacements. At present, these two sets of equations are coupled. To decouple them we first eliminate the dipole moment from Eqs. (12.6) and (12.7).

By remembering that

$$\mathbf{C} = 1 - \mathbf{aUHU} \quad (12.9)$$

we obtain

$$(1 + \mathbf{SU})\mathbf{HU\check{C}}^{-1}\mathbf{aE} + \mathbf{T}_\xi = [\mathbf{R} - (1 + \mathbf{SU})\mathbf{HUC}^{-1}(1 + \mathbf{U\check{S}})]\xi$$

$$= \mathbf{M}\xi \quad (12.10)$$

where $\mathbf{M}$ is the force-constant matrix for the dynamical problem. Now $\mathbf{T}_\xi$ contains an electrical part $(\mathbf{U}^{-1} + \mathbf{S})\mathbf{E}$ due to the interaction between the electric field generated by region I and the displacement and deformation-dipole moments in region II, $\mathbf{U}^{-1}\boldsymbol{\xi}$ and $\mathbf{\check{S}}\boldsymbol{\xi}$, respectively. It also contains a nonelectrical part $\mathbf{V}$ due to the closed-shell interactions. We therefore write

$$\mathbf{T}_\xi = (\mathbf{U}^{-1} + \mathbf{S})\mathbf{E} + \mathbf{V} \tag{12.11}$$

Equation (12.10) then becomes

$$\mathbf{M}\boldsymbol{\xi} = \mathbf{U}^{-1}(\mathbf{1} + \mathbf{US})\mathbf{\check{C}}^{-1}\mathbf{E} + \mathbf{V} \tag{12.12}$$

We now substitute back into Eq. (12.7) and obtain for the dipole moment

$$\boldsymbol{\mu} = \mathbf{C}^{-1}\mathbf{a}[\mathbf{1} + \mathbf{UH}(\mathbf{1} + \mathbf{U\check{S}})\mathbf{M}^{-1}\mathbf{U}^{-1}(\mathbf{1} + \mathbf{US})\mathbf{\check{C}}^{-1}]\mathbf{E}$$

$$+ \mathbf{C}^{-1}\mathbf{a}\mathbf{UH}(\mathbf{1} + \mathbf{U\check{S}})\mathbf{M}^{-1}\mathbf{V} \tag{12.13}$$

In order to obtain the relaxation energy at equilibrium we substitute from Eqs. (12.6) and (12.7) back into Eq. (12.4); this tells us that

$$Y_{\min} = \tfrac{1}{2}[\mathbf{\tilde{E}}(\mathbf{\check{S}U} + \mathbf{1})\mathbf{U}^{-1}\boldsymbol{\xi}_{\min} + \mathbf{\tilde{V}}\boldsymbol{\xi}_{\min} + \mathbf{\tilde{E}}\boldsymbol{\mu}_{\min}] \tag{12.14}$$

If we now substitute for $\boldsymbol{\xi}_{\min}$ and $\boldsymbol{\mu}_{\min}$ in Eq. (12.14) from Eqs. (12.12) and (12.13), we find that

$$Y_{\min} = \tfrac{1}{2}\{[\mathbf{\tilde{V}} + \mathbf{\tilde{E}}\mathbf{C}^{-1}(\mathbf{1} + \mathbf{\check{S}U})\mathbf{U}^{-1}]\mathbf{M}^{-1}[\mathbf{U}^{-1}(\mathbf{1} + \mathbf{US})\mathbf{\check{C}}^{-1}\mathbf{E} + \mathbf{V}]\}$$

$$+ \tfrac{1}{2}\mathbf{\tilde{E}}\mathbf{C}^{-1}\mathbf{a}\mathbf{E} \tag{12.15}$$

The relaxation energy, which is the change in energy when the ions are allowed to displace and polarize, is given by

$$E_R = -Y_{\min} \qquad (12.16)$$

within the zero-order approximation, when we evaluate the forces $\mathbf{E}$ and $\mathbf{V}$ at the undisplaced ion positions. This result is correct to first order in $\boldsymbol{\xi}$ and $\boldsymbol{\mu}$. If one works to higher accuracy, explicit evaluation of certain direct-space displacements is necessary and the expression for E_R loses some of its simplicity. If we are concerned only as to the validity of the zero-order approximation for first-neighbor displacements, the latter must be evaluated explicitly. This is done by subjecting the equation for the displacements to a Fourier transform. The Fourier amplitudes of the displacement field then depend on the magnitude of the first-neighbor displacement. Thus summing the Fourier series for the first-neighbor displacements yields soluble nonlinear equations for these quantities.

The allowed wave vectors $\mathbf{q}$ are restricted by the imposition of periodic boundary conditions. For the defect problem, this is equivalent to treating a superlattice of defects, one per N unit cells. Furthermore, we are at present constraining the volume of the crystal to remain constant. It follows that the displacements and dipole moments that we obtain are those appropriate to the defect super-lattice. However, we shall show that it is possible to increase our sample of $\mathbf{q}$ vectors until the displacements and dipole moments of the ions in a relatively large region about the defect become independent of the sample density. It is thus reasonable to assume that they are the same as those for a single defect in an infinite crystal.

Once the Fourier amplitudes of the displacement and dipole fields have been determined, allowing for first-neighbor relaxation, the direct-space displacements and dipole moments for any ion can be obtained by Fourier synthesis.

It should be remembered that the dipoles calculated by this procedure are the field-induced dipoles. In order to obtain the total

dipole moment on a given ion it is necessary to include the deformation-dipole amplitude, defined by the Fourier transform of Eq.
(12.4a). This equation gives the Fourier transform of the deformation-dipole moment; the direct-space dipole has to be calculated
by Fourier synthesis.

It can be seen from the foregoing analysis that lattice statics
provides a very powerful tool for the theoretical study of defects in
ionic crystals, since we have already seen in Section 6 how it is
possible to evaluate the Fourier-transformed dipole–dipole and
monopole–monopole forces in a rapidly convergent manner. We
present in Tables X and XI the results of calculations of displacements and zero-order vacancy-formation energies for the sequence of
alkali halides with the rock-salt structure. In this context it should be
noted that the relevant energy is the energy to form a Schottky pair
(i.e., a well-separated positive- and negative-ion vacancy pair). We
have used the simplest Born–Mayer model to compute the lattice
energy; thus we consider only nearest-neighbor short-range repulsive interactions.

More refined calculations have been carried out both for
Schottky defects[210] and for Frenkel defects.[211] Even without these
refinements, one can see from Table XI that the agreement between
theory and the experimental values, where the latter exist, is satisfactory and at least as good as that obtained by the Mott–Littleton
technique and its refinements.

The advantages of the present technique are twofold: (1) there is
no need for any matching and (2) it is computationally faster. The
relaxation energy in any given case was computed using an increas-
ingly dense sample of q vectors. In practice it appears that one can
extrapolate reliably from the results obtained for a sample of 64,000 q
vectors to those appropriate for an infinite sample. The reliability of
this extrapolation was checked by doing runs for a sample of 512,000
q vectors for one or two cases.

At large distances from the defect we have to consider the
problem of calculating the asymptotic forms of the displacement and

TABLE X. Displacement Components for Certain Near Neighbors About a Positive-Ion Vacancy or Negative-Ion Vacancy[a]

Neighbor			Positive-ion vacancy			Negative-ion vacancy		
L_1	L_2	L_3	ξ_x	ξ_y	ξ_z	ξ_x	ξ_y	ξ_z
Lithium fluoride								
1	0	0	+12.356	0.0	0.0	+16.057	0.0	0.0
1	1	0	−1.635	−1.635	0.0	−1.316	−1.316	0.0
1	1	1	+0.160	+0.160	+0.160	+0.186	+0.186	+0.186
2	0	0	−0.622	0.0	0.0	+1.428	0.0	0.0
1	2	0	+1.752	+1.654	0.0	+2.227	+2.355	0.0
1	1	2	−0.567	−0.567	−0.455	−0.432	−0.432	−0.432
2	2	0	−0.377	−0.377	0.0	+0.024	+0.024	0.0
2	2	1	+0.345	+0.345	+0.444	+0.561	+0.561	+0.609
3	0	0	+0.589	0.0	0.0	+0.454	0.0	0.0
2	2	2	−0.316	−0.316	−0.316	−0.124	−0.124	−0.124
Lithium chloride								
1	0	0	+7.221	0.0	0.0	+14.262	0.0	0.0
1	1	0	−0.925	−0.925	0.0	−1.319	−1.319	0.0
1	1	1	+0.324	+0.324	+0.324	+0.248	+0.248	+0.248
2	0	0	−2.522	0.0	0.0	+3.271	0.0	0.0
1	2	0	+1.366	+0.967	0.0	+1.387	+1.516	0.0
1	1	2	−0.282	−0.282	−0.155	−0.412	−0.412	−0.348
2	2	0	−0.130	−0.130	0.0	+0.015	+0.015	0.0
2	2	1	+0.303	+0.303	+0.386	+0.301	+0.301	+0.285
3	0	0	−0.387	0.0	0.0	+2.639	0.0	0.0
2	2	2	−0.100	−0.100	−0.100	−0.182	−0.182	−0.182
Lithium bromide								
1	0	0	+5.433	0.0	0.0	+13.839	0.0	0.0
1	1	0	−0.798	−0.798	0.0	−1.274	−1.274	0.0
1	1	1	+0.331	+0.331	+0.331	+0.263	+0.263	+0.263
2	0	0	−3.134	0.0	0.0	+3.861	0.0	0.0
1	2	0	+1.225	+0.779	0.0	+1.232	+1.359	0.0
1	1	2	−0.243	−0.243	−0.129	−0.385	−0.385	−0.321
2	2	0	−0.112	−0.112	0.0	+0.045	+0.045	0.0
2	2	1	+0.268	+0.268	+0.352	+0.264	+0.264	+0.236
3	0	0	−0.767	0.0	0.0	+3.028	0.0	0.0
2	2	2	−0.078	−0.078	−0.078	−0.170	−0.170	−0.170

TABLE X (*contd.*)

Neighbor			Positive-ion vacancy			Negative-ion vacancy		
L_1	L_2	L_3	ξ_x	ξ_y	ξ_z	ξ_x	ξ_y	ξ_z
Lithium iodide								
1	0	0	+3.444	0.0	0.0	+14.108	0.0	0.0
1	1	0	−0.226	−0.226	0.0	−1.222	−1.222	0.0
1	1	1	+0.303	+0.303	+0.303	+0.192	+0.192	+0.192
2	0	0	−5.370	0.0	0.0	+4.427	0.0	0.0
1	2	0	+1.188	+0.784	0.0	+1.097	+1.019	0.0
1	1	2	−0.114	−0.114	+0.038	−0.366	−0.366	−0.298
2	2	0	−0.110	−0.110	0.0	+0.063	+0.063	0.0
2	2	1	+0.277	+0.277	+0.366	+0.178	+0.178	+0.171
3	0	0	−1.783	0.0	0.0	+4.341	0.0	0.0
2	2	2	−0.024	−0.024	−0.024	−0.177	−0.177	−0.177
Sodium fluoride								
1	0	0	+11.128	0.0	0.0	+13.202	0.0	0.0
1	1	0	−1.965	−1.965	0.0	−1.465	−1.465	0.0
1	1	1	+0.441	+0.441	+0.441	+0.576	+0.576	+0.576
2	0	0	+1.012	0.0	0.0	+1.650	0.0	0.0
1	2	0	+1.191	+1.286	0.0	+1.620	+1.805	0.0
1	1	2	−0.600	−0.600	−0.677	−0.415	−0.415	−0.369
2	2	0	−0.415	−0.415	0.0	−0.050	−0.050	0.0
2	2	1	+0.305	+0.305	+0.277	+0.523	+0.523	+0.443
3	0	0	+1.435	0.0	0.0	+0.776	0.0	0.0
2	2	2	−0.347	−0.347	−0.347	−0.145	−0.145	−0.145
Sodium chloride								
1	0	0	+7.625	0.0	0.0	+12.527	0.0	0.0
1	1	0	−1.517	−1.517	0.0	−1.367	−1.367	0.0
1	1	1	+0.429	+0.429	+0.429	+0.512	+0.512	+0.512
2	0	0	−0.301	0.0	0.0	+3.016	0.0	0.0
1	2	0	+1.052	+0.802	0.0	+1.232	+1.467	0.0
1	1	2	−0.422	−0.422	−0.454	−0.376	−0.376	−0.330
2	2	0	−0.230	−0.230	0.0	−0.010	−0.010	0.0
2	2	1	+0.249	+0.249	+0.267	+0.374	+0.374	+0.293
3	0	0	+0.587	0.0	0.0	+1.820	0.0	0.0
2	2	2	−0.200	−0.200	−0.200	−0.150	−0.150	−0.150

TABLE X (*contd.*)

Neighbor			Positive-ion vacancy			, Negative-ion vacancy		
L_1	L_2	L_3	ξ_x	ξ_y	ξ_z	ξ_x	ξ_y	ξ_z

Sodium bromide

1	0	0	+6.657	0.0	0.0	+12.639	0.0	0.0
1	1	0	−1.352	−1.352	0.0	−1.311	−1.311	0.0
1	1	1	+0.398	+0.398	+0.398	+0.458	+0.458	+0.458
2	0	0	−0.946	0.0	0.0	+3.353	0.0	0.0
1	2	0	+1.051	+0.712	0.0	+1.192	+1.417	0.0
1	1	2	−0.376	−0.376	−0.380	−0.364	−0.364	−0.306
2	2	0	−0.189	−0.189	0.0	+0.018	+0.018	0.0
2	2	1	+0.235	+0.235	+0.273	+0.340	+0.340	+0.270
3	0	0	+0.245	0.0	0.0	+2.100	0.0	0.0
2	2	2	−0.166	−0.166	−0.166	−0.146	−0.146	−0.146

Sodium iodide

1	0	0	+5.338	0.0	0.0	+12.175	0.0	0.0
1	1	0	−0.893	−0.893	0.0	−1.272	−1.272	0.0
1	1	1	+0.380	+0.380	+0.380	+0.363	+0.363	+0.363
2	0	0	−2.068	0.0	0.0	+3.785	0.0	0.0
1	2	0	+1.039	+0.638	0.0	+0.930	+1.109	0.0
1	1	2	−0.250	−0.250	−0.190	−0.356	−0.356	−0.303
2	2	0	−0.084	−0.084	0.0	−0.031	−0.031	0.0
2	2	1	+0.241	+0.241	+0.288	+0.231	+0.231	+0.175
3	0	0	−0.381	0.0	0.0	+3.092	0.0	0.0
2	2	2	−0.079	−0.079	−0.079	−0.172	−0.172	−0.172

Potassium fluoride

1	0	0	+11.312	0.0	0.0	+11.294	0.0	0.0
1	1	0	−1.723	−1.723	0.0	−1.123	−1.123	0.0
1	1	1	+0.387	+0.387	+0.387	+0.524	+0.524	+0.524
2	0	0	+1.508	0.0	0.0	+0.377	0.0	0.0
1	2	0	+1.002	+1.121	0.0	+1.464	+1.414	0.0
1	1	2	−0.515	−0.515	−0.562	−0.309	−0.309	−0.231
2	2	0	−0.336	−0.336	0.0	−0.018	−0.018	0.0
2	2	1	+0.244	+0.244	+0.206	+0.440	+0.440	+0.400
3	0	0	+2.321	0.0	0.0	+0.663	0.0	0.0
2	2	2	−0.300	−0.300	−0.300	−0.900	−0.090	−0.090

TABLE X (contd.)

Neighbor			Positive-ion vacancy			Negative-ion vacancy		
L_1	L_2	L_3	ξ_x	ξ_y	ξ_z	ξ_x	ξ_y	ξ_z

Potassium chloride

1	0	0	+8.390	0.0	0.0	+11.080	0.0	0.0
1	1	0	−1.615	−1.615	0.0	−1.240	−1.240	0.0
1	1	1	+0.437	+0.437	+0.437	+0.566	+0.566	+0.566
2	0	0	+0.805	0.0	0.0	+2.093	0.0	0.0
1	2	0	+0.879	+0.780	0.0	+1.175	+1.273	0.0
1	1	2	−0.438	−0.438	−0.501	−0.318	−0.318	−0.278
2	2	0	−0.265	−0.265	0.0	+0.004	+0.004	0.0
2	2	1	+0.216	+0.216	+0.203	+0.376	+0.376	+0.299
3	0	0	+1.409	0.0	0.0	+1.389	0.0	0.0
2	2	2	−0.231	−0.231	−0.231	−0.109	−0.109	−0.109

Potassium bromide

1	0	0	+7.508	0.0	0.0	+11.081	0.0	0.0
1	1	0	−1.518	−1.518	0.0	−1.220	−1.220	0.0
1	1	1	+0.423	+0.423	+0.423	+0.546	+0.546	+0.546
2	0	0	+0.407	0.0	0.0	+2.500	0.0	0.0
1	2	0	+0.865	+0.677	0.0	+1.106	+1.230	0.0
1	1	2	−0.405	−0.405	−0.458	−0.312	−0.312	−0.265
2	2	0	−0.228	−0.228	0.0	+0.015	+0.015	0.0
2	2	1	+0.204	+0.204	+0.204	+0.349	+0.349	+0.272
3	0	0	+1.116	0.0	0.0	+1.634	0.0	0.0
2	2	2	−0.203	−0.203	−0.203	−0.110	−0.110	−0.110

Potassium iodide

1	0	0	+6.329	0.0	0.0	+11.103	0.0	0.0
1	1	0	−1.358	−1.358	0.0	−1.185	−1.185	0.0
1	1	1	+0.400	+0.400	+0.400	+0.506	+0.506	+0.506
2	0	0	−0.227	0.0	0.0	+3.008	0.0	0.0
1	2	0	+0.854	+0.558	0.0	+1.015	+1.166	0.0
1	1	2	−0.357	−0.357	−0.390	−0.304	−0.304	−0.247
2	2	0	−0.177	−0.177	0.0	+0.025	+0.025	0.0
2	2	1	+0.191	+0.191	+0.209	+0.309	+0.309	+0.236
3	0	0	+0.687	0.0	0.0	+1.999	0.0	0.0
2	2	2	−0.164	−0.164	−0.164	−0.113	−0.113	−0.113

TABLE X (contd.)

Neighbor			Positive-ion vacancy			Negative-ion vacancy		
L_1	L_2	L_3	ξ_x	ξ_y	ξ_z	ξ_x	ξ_y	ξ_z
Rubidium fluoride								
1	0	0	+11.754	0.0	0.0	+10.764	0.0	0.0
1	1	0	−1.485	−1.485	0.0	−0.923	−0.923	0.0
1	1	1	+0.336	+0.336	+0.336	+0.445	+0.445	+0.445
2	0	0	+1.343	0.0	0.0	−0.462	0.0	0.0
1	2	0	+1.029	+1.082	0.0	+1.427	+1.278	0.0
1	1	2	−0.451	−0.451	−0.461	−0.267	−0.267	−0.164
2	2	0	−0.260	−0.260	0.0	−0.036	−0.036	0.0
2	2	1	+0.234	+0.234	+0.204	+0.390	+0.390	+0.382
3	0	0	+2.686	0.0	0.0	+0.718	0.0	0.0
2	2	2	−0.255	−0.255	−0.255	−0.081	−0.081	−0.081
Rubidium chloride								
1	0	0	+8.620	0.0	0.0	+10.395	0.0	0.0
1	1	0	−1.542	−1.542	0.0	−1.180	−1.180	0.0
1	1	1	+0.435	+0.435	+0.435	+0.550	+0.550	+0.550
2	0	0	+1.056	0.0	0.0	+1.683	0.0	0.0
1	2	0	+0.834	+0.777	0.0	+1.110	+1.147	0.0
1	1	2	−0.413	−0.413	−0.466	−0.299	−0.299	−0.265
2	2	0	−0.242	−0.242	0.0	−0.008	−0.008	0.0
2	2	1	+0.212	+0.212	+0.188	+0.350	+0.350	+0.282
3	0	0	+1.660	0.0	0.0	+1.352	0.0	0.0
2	2	2	−0.218	−0.218	−0.218	−0.104	−0.104	−0.104
Rubidium bromide								
1	0	0	+7.860	0.0	0.0	+10.615	0.0	0.0
1	1	0	−1.517	−1.517	0.0	−1.162	−1.162	0.0
1	1	1	+0.417	+0.417	+0.417	+0.543	+0.543	+0.543
2	0	0	+0.729	0.0	0.0	+2.100	0.0	0.0
1	2	0	+0.825	+0.687	0.0	+1.101	+1.166	0.0
1	1	2	−0.405	−0.405	−0.456	−0.293	−0.293	−0.247
2	2	0	−0.232	−0.232	0.0	+0.021	+0.021	0.0
2	2	1	+0.195	+0.195	+0.188	+0.345	+0.345	+0.276
3	0	0	+1.403	0.0	0.0	+1.478	0.0	0.0
2	2	2	−0.209	−0.209	−0.209	−0.097	−0.097	−0.097

TABLE X (*contd.*)

Neighbor			Positive-ion vacancy			Negative-ion vacancy		
L_1	L_2	L_3	ξ_x	ξ_y	ξ_z	ξ_x	ξ_y	ξ_z

Rubidium iodide

L_1	L_2	L_3	ξ_x	ξ_y	ξ_z	ξ_x	ξ_y	ξ_z
1	0	0	+6.770	0.0	0.0	+10.628	0.0	0.0
1	1	0	−1.408	−1.408	0.0	−1.155	−1.155	0.0
1	1	1	+0.406	+0.406	+0.406	+0.524	+0.524	+0.524
2	0	0	+0.237	0.0	0.0	+2.642	0.0	0.0
1	2	0	+0.808	+0.571	0.0	+1.008	+1.117	0.0
1	1	2	−0.367	−0.367	−0.409	−0.289	−0.289	−0.237
2	2	0	−0.192	−0.192	0.0	+0.025	+0.025	0.0
2	2	1	+0.185	+0.185	+0.191	+0.314	+0.314	+0.242
3	0	0	+1.008	0.0	0.0	+1.802	0.0	0.0
2	2	2	−0.177	−0.177	−0.177	−0.102	−0.102	−0.102

Cesium fluoride

L_1	L_2	L_3	ξ_x	ξ_y	ξ_z	ξ_x	ξ_y	ξ_z
1	0	0	+10.067	0.0	0.0	+8.614	0.0	0.0
1	1	0	−1.735	−1.735	0.0	−0.690	−0.690	0.0
1	1	1	+0.319	+0.319	+0.319	+0.534	+0.534	+0.534
2	0	0	+2.773	0.0	0.0	−0.491	0.0	0.0
1	2	0	+0.548	+0.690	0.0	+1.295	+1.075	0.0
1	1	2	−0.488	−0.488	−0.546	−0.174	−0.174	−0.056
2	2	0	−0.387	−0.387	0.0	+0.092	+0.092	0.0
2	2	1	+0.113	+0.113	+0.082	+0.397	+0.397	+0.376
3	0	0	+3.553	0.0	0.0	−0.033	0.0	0.0
2	2	2	−0.310	−0.310	−0.310	+0.0001	+0.0001	+0.0001

[a] Values are in percent of the near-neighbor distance in the perfect lattice. A positive value indicates outward relaxation $[\mathbf{r}\binom{l}{k} = r_0(L_1, L_2, L_3)]$.

polarization fields. This is valuable in order to establish what the asymptotic expressions used in region II in a Mott–Littleton calculation should be. This particular problem was first discussed by Hardy and Lidiard.[96] The crucial point is that in the asymptotic limit the displacements and dipole moments are determined only by their Fourier amplitudes in the region about the origin of reciprocal space. This is because the Fourier series oscillate very rapidly as the

TABLE XI. Calculated Energies[a] Needed to Remove Positive and Negative Ions to Infinity (E_+ and E_-), Lattice Energies (E_L), and Values of Schottky-Pair Formation Energy (E_S)

Parameter	LiF	LiCl	LiBr	LiI	NaF	NaCl	NaBr	NaI	KF	KCl	KBr	KI	RbF	RbCl	RbBr	RbI	CsF
E_+	6 012	4 640	4 290	3 739	5 929	4 715	4 332	3 907	5 326	4 500	4 233	3 867	5 026	4 388	4 142	3 811	5 020
E_-	6 331	5 347	5 068	4 636	6 007	5 126	4 822	4 523	5 141	4 633	4 453	4 190	4 773	4 439	4 271	4 042	4 538
E_{tot}	12 343	9 987	9 358	8 375	11 936	9 841	9 154	8 430	10 467	9 133	8 686	8 057	9 799	8 827	8 413	7 853	9 558
E_L	−10 654	−8 607	−8 077	−7 436	−9 537	−7 937	−7 497	−6 989	−8 397	−7 198	−6 875	−6 436	−7 995	−6 927	−6 643	−6 215	−7 621
E_S	1 689	1 380	1 281	0 939	2 399	1 904	1 657	1 441	2 070	1 935	1 811	1 621	1 804	1 900	1 770	1 638	1 937
E_S (BL)[b]	1 797	1 075	0 864	—	2 491	1 841	1 631	1 347	2 178	2 052	1 920	1 719	1 981	—	1 937	1 798	—
E_S (expt)[c]	2 68	2 12	1 80	1 34	—	2 02	1 68	—	2 64[d]	2 22	2 53	1 60	—	2.0[e]	—	—	—
						2 09	1 74			2 30	1 99	1 58					
						2 12					2 00	1 56					

[a] In electron volts

[b] I M Boswarra and A B Lidiard, *Phil Mag* **16**, 805 (1967), Table 4

[c] Experimental values tabulated by P V Sastry and B G Mulimann, *Phil Mag* **20**, 166 (1969)

[d] P Suptitz and J Teltow, *Phys Status Solidi* **23**, 9 (1967)

[e] J B Holt, H G Sockel, and H Schmalzried, *J Am Cer Soc* **52**, 376 (1969)

wave vector $\mathbf{q}$ varies. Consequently the contributions from the outer parts of the Brillouin zone cancel. Thus one can use for the Fourier components the asymptotic expressions valid in the limit $\mathbf{q} \to 0$.

Let us first consider the equation [Eq. (12.10)] for the Fourier amplitudes of the displacements. This we can express in the following form:

$$\mathbf{MQ} = \mathbf{U}^{-1}(1 + \mathbf{US})\check{\mathbf{C}}^{-1}\mathbf{E} + \mathbf{V} \equiv \mathbf{F} \tag{12.17}$$

or in block form:

$$\begin{bmatrix} \mathbf{M}_{11} & \mathbf{M}_{12} \\ \mathbf{M}_{12} & \mathbf{M}_{11} \end{bmatrix}\begin{bmatrix} \mathbf{Q}_1 \\ \mathbf{Q}_2 \end{bmatrix} = \begin{bmatrix} \mathbf{F}_1 \\ \mathbf{F}_2 \end{bmatrix} \tag{12.18}$$

where the blocks are 3×3, or three-component, and the suffixes 1 and 2 refer to cations and anions, respectively; furthermore, $\mathbf{M}$ is symmetric and $\mathbf{M}_{11} = \mathbf{M}_{22}$. Equation (12.18) is such that we have two separate equations, one for $\mathbf{Q}_1 + \mathbf{Q}_2$ and one for $\mathbf{Q}_1 - \mathbf{Q}_2$. The sum corresponds to a strain field, and the difference to a polarization field.

Thus we have for the asymptotic strain field and polarization the following equations:

$$(\mathbf{M}_{11} + \mathbf{M}_{12})(\mathbf{Q}_1 + \mathbf{Q}_2) = \mathbf{F}_1 + \mathbf{F}_2 \tag{12.19}$$

and

$$(\mathbf{M}_{11} - \mathbf{M}_{12})(\mathbf{Q}_1 - \mathbf{Q}_2) = \mathbf{F}_1 - \mathbf{F}_2 \tag{12.20}$$

in the limit $\mathbf{q} \to 0$.

The main problem is to determine

$$\mathbf{F}^c \equiv \mathbf{U}^{-1}(1 + \mathbf{US})\tilde{\mathbf{C}}^{-1}\mathbf{E} \tag{12.21}$$

in the limit $\mathbf{q} \to 0$. We introduce

$$\mathbf{Z} = \check{\mathbf{C}}^{-1}\mathbf{E} \tag{12.22}$$

Also, when the negative ions alone are deformable,

$$\mathbf{S} = \begin{bmatrix} 0 & \mathbf{S}_{12} \\ 0 & \mathbf{S}_{22} \end{bmatrix}$$

and $\mathbf{S}_{12} \to -\mathbf{S}_{22}$ as $\mathbf{q} \to 0$. Thus

$$\mathbf{F}_1^c + \mathbf{F}_2^c = e_1\mathbf{Z}_1 + e_2\mathbf{Z}_2 \tag{12.23}$$

and

$$\mathbf{F}_1^c - \mathbf{F}_2^c = e_1\mathbf{Z}_1 - e_2\mathbf{Z}_2 + 2\mathbf{S}_{12}\mathbf{Z}_2 \tag{12.24}$$

in the limit $\mathbf{q} \to 0$. However,

$$\mathbf{CZ} = \mathbf{E}$$

and hence

$$\begin{bmatrix} \left(1 - \dfrac{\alpha_1}{e_1^2}\mathbf{H}_{11}\right)\mathbf{Z}_1 - \dfrac{\alpha_2}{e_1 e_2}\mathbf{H}_{12}\mathbf{Z}_2 \\[2mm] -\dfrac{\alpha_1}{e_1 e_2}\mathbf{H}_{12}\mathbf{Z}_1 + \left(1 - \dfrac{\alpha_2}{e_2^2}\mathbf{H}_{22}\right)\mathbf{Z}_2 \end{bmatrix} = \begin{bmatrix} \mathbf{E}_1 \\[2mm] \mathbf{E}_2 \end{bmatrix} \tag{12.25}$$

Now $\mathbf{H}_{22} = \mathbf{H}_{11}$ and $\mathbf{H}_{11} \to -\mathbf{H}_{12}$ as $\mathbf{q} \to 0$, since $e_1 = -e_2$. Thus Eqs. (12.23) and (12.25) give

$$\mathbf{F}_1^c + \mathbf{F}_2^c = e(\mathbf{Z}_1 - \mathbf{Z}_2) = e(\mathbf{E}_1 - \mathbf{E}_2) \tag{12.26}$$

The quantity $\mathbf{F}_1 + \mathbf{F}_2$ is obtained by adding to this the contribution from the short-range forces $\mathbf{V}_1 + \mathbf{V}_2$.

The quantity $\mathbf{F}_1 - \mathbf{F}_2$ is needed for the polarization equation, Eq. (12.20). Equation (12.25) gives

$$\mathbf{E}_1 + \mathbf{E}_2 = \mathbf{Z}_1 + \mathbf{Z}_2 - (2\mathbf{H}_{11}/e^2)(\alpha_1 \mathbf{Z}_1 + \alpha_2 \mathbf{Z}_2) \qquad (12.27)$$

As $\mathbf{q}$ goes to zero, $\mathbf{M}_{11} - \mathbf{M}_{12}$ becomes independent of $\mathbf{q}$. The dominant term in $\mathbf{F}_1 - \mathbf{F}_2$ as $\mathbf{q}$ goes to zero is the leading term in $\mathbf{F}_1^c - \mathbf{F}_2^c$, which is $\sim \mathbf{q}/q^2$. To the same order in $\mathbf{q}$, $\mathbf{E}_1 = \mathbf{E}_2$, and Eq. (12.26) gives $\mathbf{Z}_1 = \mathbf{Z}_2 \equiv \mathbf{Z}$. Thus Eq. (12.27) reduces to

$$\mathbf{E}_1 + \mathbf{E}_2 = 2\left(1 - \frac{\alpha_1 + \alpha_2}{e^2}\mathbf{H}_{11}\right)\mathbf{Z} \qquad (12.28)$$

Hence

$$\mathbf{F}_1^c - \mathbf{F}_2^c = e\left(1 + \frac{\mathbf{S}_{12}}{e}\right)\left(1 - \frac{\alpha_1 + \alpha_2}{e^2}\mathbf{H}_{11}\right)^{-1}(\mathbf{E}_1 + \mathbf{E}_2) \quad (12.29)$$

To determine the polarization field we must back-transform the limiting form of $\mathbf{Q}_1 - \mathbf{Q}_2$ as $\mathbf{q}$ approaches zero. In this limit

$$\mathbf{E}_1 = \mathbf{E}_2 = \frac{2ize}{v_a}\frac{\mathbf{q}}{q^2}$$

Thus, from Eq. (12.29), we have

$$\frac{\mathbf{F}_1^c - \mathbf{F}_2^c}{2} = ze^2\left(1 + \frac{\mathbf{S}_{12}}{e}\right)\left(1 - \frac{\alpha_1 + \alpha_2}{e^2}\mathbf{H}_{11}\right)^{-1}\frac{2i\mathbf{q}}{v_a q^2} \qquad (12.30)$$

In the limit $\mathbf{q} \to 0$, $\mathbf{M}_{11} - \mathbf{M}_{12}$ becomes independent of $\mathbf{q}$; hence $\mathbf{Q}_1 - \mathbf{Q}_2$ is determined by $\mathbf{F}_1^c - \mathbf{F}_2^c$ alone since the remaining part of $\mathbf{F}_1 - \mathbf{F}_2$ is only $\sim \mathbf{q}$.

Further it can be shown (see Section 6) that, as $\mathbf{q}$ approaches zero,

$$H_{11}^{\alpha\beta} \to -\frac{4\pi e^2}{v_a}\left(\frac{q_\alpha q_\beta}{q^2} - \frac{\delta_{\alpha\beta}}{3}\right) \tag{12.31}$$

whereas

$$S_{12}^{\alpha\beta} \to S\delta_{\alpha\beta} = (e^* - e)\delta_{\alpha\beta}$$

where e^* is the Szigeti effective charge.

Hence, if we transform to polar coordinates in $\mathbf{q}$ space with the polar axis along $\mathbf{q}$, we obtain

$$\frac{\mathbf{F}_1^c - \mathbf{F}_2^c}{2} = \frac{zee^*}{1 + (8\pi/3v_a)(\alpha_1 + \alpha_2)}\frac{2i\mathbf{q}}{v_a q^2}$$

$$= zee^*\left(\frac{\varepsilon_\infty + 2}{3\varepsilon_\infty}\right)\frac{2i\mathbf{q}}{v_a q^2} \tag{12.32}$$

But $\mathbf{M}_{11} - \mathbf{M}_{12}$ is also diagonalized by the same transformation and becomes

$$2N\bar{M} \times \begin{bmatrix} \omega_l^2 & 0 & 0 \\ 0 & \omega_0^2 & 0 \\ 0 & 0 & \omega_0^2 \end{bmatrix} \tag{12.33}$$

where ω_l and ω_0 are the limiting frequencies, as $\mathbf{q}$ approaches zero, of the longitudinal and transverse optical vibrations.

Thus the Fourier components of the polarization field are

$$\frac{\mathbf{Q}_1 - \mathbf{Q}_2}{2} = \frac{2ize^*e}{2Nv_a\bar{M}\omega_l^2}\left(\frac{\varepsilon_\infty + 2}{3\varepsilon_\infty}\right)\frac{\mathbf{q}}{q^2} \tag{12.34}$$

which on back-transforming gives

$$\frac{\xi_1 - \xi_2}{2} = \frac{zee^*}{2\bar{M}\omega_i^2}\left(\frac{\varepsilon_\infty + 2}{3\varepsilon_\infty}\right)\frac{\mathbf{r}}{r^3}$$

$$= \frac{ee^*}{2K}\left(\frac{z\mathbf{r}}{r^3}\right)\frac{\varepsilon_0 + 2}{3\varepsilon_0} \tag{12.35}$$

To obtain the result shown in Eq. (12.35) we have used the Lyddane–Sachs–Teller relation and the first Szigeti relation[44] to introduce the parameter

$$K = 6r_0/\beta$$

This is to be compared with the corresponding Mott–Littleton[198] expression, which can be cast in the form

$$\frac{\xi_1 - \xi_2}{2} = \frac{z\mathbf{r}}{r^3}\left(\frac{ee}{2K}\right)\frac{\varepsilon_0 + 2}{3\varepsilon_0} \tag{12.36}$$

The only difference between the two is that in our Eq. (12.35) the Szigeti effective charge replaces the full electronic charge.

Similarly, we can manipulate the equations for the strain field to obtain

$$(F_1 + F_2)_\alpha = \sum_{lk}\left\{-iq_\alpha r_\alpha\binom{l}{k}\left[\partial/\partial\xi\binom{l}{k}_\alpha\right]\phi\left[\left|\mathbf{r}\binom{l}{k} + \xi\binom{l}{k}\right|\right]\right\} \tag{12.37}$$

where ϕ is the total defect-lattice interaction. This result can be put in direct correspondence with that of continuum-elasticity theory since it can be shown that the transformed elastic response of such a continuum subject to a body force distribution $-G \nabla \delta(\mathbf{r})$ is given by

$$\sum_\beta (M_{11} + M_{12})_{\alpha\beta}(Q_1 + Q_2)_\beta = -iGq_\alpha \qquad (12.38)$$

when the elements of $\mathbf{M}_{11} + \mathbf{M}_{12}$ are expressed in terms of the elastic constants.

Thus we have an expression for the transform of the strain-field displacement that is uniquely defined in terms of microscopic quantities. One can demonstrate an interesting result at this point. From Eqs. (12.37) and (12.38) the strength parameter is given by

$$G = \sum_{lk} r\binom{l}{k}_\alpha \left[\partial \Big/ \partial \xi \binom{l}{k}_\alpha \right] \phi \left[\left| \mathbf{r} \binom{l}{k} + \boldsymbol{\xi} \binom{l}{k} \right| \right] \qquad (12.39)$$

If we expand the derivative as a power series in the displacements of the ions and restrict our short-range forces to act between first neighbors only, the leading term in G (G_0) is zero because of the equilibrium condition for the perfect lattice. This is true for either a positive or a negative ion vacancy, but Faux and Lidiard[202] have pointed out that, if the short-range forces are extended to second neighbors, then it may only be asserted that the *net* strength for a Schottky pair is zero. Thus we see that the elastic strength of either a positive or a negative ion vacancy is mainly associated with the relaxation. These authors[202] also pointed out that there is an additional term to that given by Hardy and Lidiard[96]—a term that arises from dipole field-gradient forces. If one proceeds to include the effect of deformation dipoles, these results have to be modified still further.[212]

However, it would appear that the values of G estimated by Schulze[212] and by Faux and Lidiard[202] differ significantly from those estimated by Brauer[204] and from earlier values obtained by Lidiard and Boswarva.[203] This is an important result for two reasons:

1. The use of an improved value of G should increase the reliability of Mott–Littleton calculations.
2. It now appears possible to make direct experimental measurements of the elastic strengths of defects.[213–216]

The lattice-statics approach can obviously be applied to other classes of defect, and such an application has already been made, as mentioned earlier, to the study of Frenkel[211] defects in alkali halides. Moreover, one could apply the technique quite readily to substitutional divalent impurities. One should bear in mind that our calculations are based on the harmonic approximation to the lattice potential energy. However, anharmonic corrections can be built in as additional external forces between ions that undergo large relative displacements.[217]

In principle the method also offers a means of examining the interaction between an itinerant electron and the crystal lattice. This is a more complicated problem because of the variable position of such an electron. As a consequence, the displacement and polarization fields will not have any inherent symmetry.[218]

There is one final point that requires clarification. The question is whether or not the elastic strength parameter, which we have calculated earlier, is the same as the parameter that determines the volume change of the crystal when we relax the earlier constraint that the supercell volume be fixed. Eshelby[219] showed that the elastic-continuum theory predicts that the volume change of a crystal should be given by $\Delta V = G\beta$. Subsequently, Hardy[220] demonstrated that this result also held for lattice theory. However, it should be verified that this conclusion is still valid for a charged defect in an ionic crystal.

Previously, we wrote the energy of the imperfect crystal as the sum of two terms, X and Y. These depend implicitly on the lattice parameter r_0. Thus we obtained equilibrium dipole moments and displacements that are functions of r_0. These we can substitute back into W and give W as a function of r_0 alone. The equilibrium value of r_0 is then obtained by setting $dW/dr_0 = 0$.

However, by the equilibrium condition this is equivalent to

$$\frac{\partial W}{\partial r_0} = \frac{\partial X}{\partial r_0} + \frac{\partial Y}{\partial r_0} = 0 \tag{12.40}$$

From the definition of Y, the term $\partial Y/\partial r_0$ is the compressibility of the perfect but distorted crystal, which within the harmonic approximation is the same as that of the perfect crystal.

If, as we assume, there are only two-body forces present, then X is given by

$$X = -\sum_{lk} \phi\left[\left|\mathbf{r}\binom{l}{k} + \boldsymbol{\xi}\binom{l}{k}\right|\right]$$

$$\frac{\partial X}{\partial r_0} = -\sum_{\substack{lk \\ \alpha}} \left[\partial/\partial r\binom{l}{k}_\alpha\right] \phi\left[\left|\mathbf{r}\binom{l}{k} + \boldsymbol{\xi}\binom{l}{k}\right|\right]\left[r\binom{l}{k}_\alpha/r_0\right]$$

$$= -\frac{3G}{r_0} \tag{12.41}$$

Thus, from Eq. (12.40), the equilibrium value of r_0 is given by

$$r_0 - a = \beta G r_0/3 N v_a \tag{12.42}$$

where a is the perfect-lattice spacing. If we use the result

$$\partial Y/\partial r_0 = 9(r_0 - a)N v_a/\beta r_0^2 \tag{12.43}$$

and the fact that the volume dilatation is

$$\Delta V = 3 N v_a (r_0 - a)/r_0 \tag{12.44}$$

and combine Eqs. (12.41)–(12.44), we find that

$$\Delta V = \beta G$$

Thus we have demonstrated that there is complete correspondence between the strength value obtained from the lattice equations and that obtained by computing the dilatation of the crystal. If we include the various dipolar contributions to G, the basic result continues to hold.

Applications of the method of lattice statics to line defects have been made by Maradudin[221] and Boyer and Hardy[222,223] for screw dislocations in sodium chloride and various metals. This class of problem is somewhat more complicated since it is necessary to find some way of simulating the effect of the screw dislocation. The technique used was to put a branch cut in the displacement field across some arbitrary plane containing the axis of the dislocation. This particular type of problem will not be discussed further since it extends outside the scope of this review.

This concludes our review of the application of the lattice-statics method to defects in alkali halide crystals. Although we have used the deformation-dipole model to treat this problem, it is evident that the same technique could be used with various types of shell model. However, if a shell model were to be used, it would be essential that it be properly matched to the dielectric constants of the perfect lattice. This condition is much more important than the requirement that the shell model give the best possible fit to the measured phonon-dispersion curves.

It should be apparent that there is much scope for extending and refining the basic ideas we have outlined. It is our feeling that the lattice-statics approach can be made at least as reliable as the most complex Mott–Littleton type of calculation and that it is computationally more efficient.

VII

Conclusions

In this book we have presented a discussion of the lattice dynamics and statics of alkali halide crystals from the viewpoint of making detailed and extensive predictions of many properties that are directly and indirectly related to the basic lattice dynamics. We have laid strong emphasis on calculations that have been based on realistic phenomenological models, in particular the deformation-dipole model. In spite of the progress that has been made in formulating lattice-dynamical theory on a more fundamental basis, it would appear that calculations of the type we have described will be necessary for some considerable time in the future.

A general criterion of the usefulness of any model is the degree to which it enables one to calculate all the normal-mode frequencies and eigenvectors from the minimum amount of input. It thus seems to us undesirable to use information obtained from the measured phonon-dispersion curves, except to a very limited degree. Such data should be used as the most direct test of the reliability of the predictions of any given model. Furthermore, many calculations we have described require information about the eigenvectors and eigenfrequencies for the modes throughout the whole zone. It is difficult to assess how well a model whose parameters have been fitted to the measured dispersion curves will reproduce either set of data (eigenfrequencies or eigenvectors). This reservation is strongest as far as the eigenvectors are concerned since neutron-scattering measurements have not been carried out to the point of providing any reliable information with regard to these quantities. Leigh, Szigeti, and Tewary[224] and Cochran[225] have shown that an infinite number of models can be devised to fit a given set of measured phonon frequencies. However, a given pair of models would in general predict very different eigenvectors.

Given a reasonable model, one can proceed to make the whole sequence of calculations described in Sections 7–12 of this book. For the majority of these calculations specific knowledge of the eigenvectors is prerequisite. We feel that it is important to present, as we have done in this book, a unified description of the numerical calculations that have been made, by others and by ourselves, since, as far as we know, no such presentation has been made elsewhere. Other workers have obviously made many separate calculations, usually in connection with the interpretation of specific experiments.

There is one final point that we should like to make concerning the comparison of lattice-dynamics calculations with band-structure calculations. The latter have been the subject of a vast amount of numerical computation, which started in the early 1950s and is still being actively pursued. However, lattice-dynamics studies have not received anything like the same amount of computational effort, and we feel that the type of work described in this review merits the same attention as that accorded the study of electronic energy bands in solids. As regards a direct comparison between the problems associated with these two types of study, one can say that calculations of electronic band structure and hence electronic densities of states require much more computer time to obtain the same degree of precision as phonon-dispersion curves and phonon densities of states. Moreover, the calculation of phonon eigenfrequencies and eigenvectors of the type we have described involves many fewer parameters than those required for a band-structure calculation. Thus it is now possible to calculate definitive spectral functions (e.g., weighted densities of states) without using inordinately large amounts of machine time.

References

1. M. BORN AND T. VON KÁRMÁN, *Physik. Z.* **13**, 297 (1912).
2. M. BORN AND T. VON KÁRMÁN, *Physik. Z.* **14**, 15 (1913).
3. M. BORN, *Dynamik der Kristallgitter*, Teubner, Leipzig (1915).
4. M. BORN, *Atomtheorie des festen Zustandes*, Teubner, Leipzig (1923).
5. W. A. HARRISON, *Pseudopotentials in the Theory of Metals*, Benjamin, New York (1966).
6. S. K. JOSHI AND A. K. RAJAGOPAL, in *Solid State Physics* (F. Seitz, D. Turnbull, and H. Ehrenreich, eds.), Vol. 22, p. 159. Academic Press, New York (1968).
7. R. M. MARTIN, *Phys. Rev.* **186**, 871 (1969).
8. N. S. GILLIS, *Phys. Rev.* **B3**, 1482 (1971).
9. T. TOYA, *J. Res. Inst. Catalysis Hokkaido Univ.* **6**, 161, 183 (1958).
10. W. COCHRAN, *Proc. Roy. Soc. (London)* **A276**, 308 (1963).
11. S. H. VOSKO, R. TAYLOR, AND G. H. KEECH, *Can. J. Phys.* **43**, 1187 (1965).
12. V. K. TEWARY AND R. BULLOUGH, *J. Phys. F. (Metal Physics)* **1**, 554 (1971).
13. J. R. HARDY AND J. W. FLOCKEN, *CRC Crit. Rev. Solid State Sci.* **1**, 605 (1971).
14. W. FRIEDRICH, P. KNIPPING, AND M. VON LAUE, *Sitzungsber. Math.–Phys. Kl. Bayer. Akad. Wiss.*, 303 (1912).
15. M. VON LAUE, *Sitzungsber. Math.–Phys. Kl. Bayer. Akad. Wiss.*, 363 (1912).
16. W. FRIEDRICH, P. KNIPPING, AND M. VON LAUE, *Ann. Physik (Leipzig)* **41**, 971 (1913).
17. M. P. TOSI, in *Solid State Physics* (F. Seitz and D. Turnbull, eds.), Vol. 16, p. 1. Academic Press, New York (1964).
18. M. BORN AND J. E. MAYER, *Z. Physik* **75**, 1 (1932).
19. M. L. HUGGINS AND J. E. MAYER, *J. Chem. Phys.* **1**, 643 (1933).
20. M. L. HUGGINS, *J. Chem. Phys.* **5**, 143 (1937); **15**, 212 (1947).

21. M. BORN AND M. GOEPPERT-MAYER, in *Handbuch der Physik*, Vol. 24, Part 2. Springer-Verlag, Berlin (1933).

22. P. O. LÖWDIN, "A Theoretical Investigation into Some Properties of Ionic Crystals," Ph.D. thesis, Uppsala University, Uppsala, Sweden. Almquist and Wiksells, Uppsala (1948).

23. E. W. KELLERMANN, *Phil. Trans. Roy. Soc. (London)* **A238**, 513 (1940).

24. P. P. EWALD, dissertation, Munich (1912).

25. P. P. EWALD, *Nachr. Ges. Wiss. Göttingen, Jahresber. Math.–Phys. Kl. Fachgrup. I*, 55 (1938).

26. K. HUANG, *Proc. Roy. Soc. (London)* **A208**, 352 (1951).

27. K. HUANG, *Nature* **167**, 779 (1951).

28. H. FRÖHLICH, *Adv. Phys.* **3**, 325 (1954).

29. A. A. MARADUDIN, E. W. MONTROLL, G. H. WEISS, AND I. P. IPATOVA, *Theory of Lattice Dynamics in the Harmonic Approximation*, 2nd ed. Academic Press, New York (1971).

30. W. LEDERMANN, *Proc. Roy. Soc. (London)* **A182**, 362 (1944).

31. M. BORN AND K. HUANG, *Dynamical Theory of Crystal Lattices*, Appendix IV, p. 391. Clarendon Press, Oxford (1954).

32. L. P. BOUCKAERDT, R. SMOLUCHOWSKI, AND E. WIGNER, *Phys. Rev.* **50**, 58 (1936).

33. M. BORN AND K. HUANG, *Dynamical Theory of Crystal Lattices*, p. 293 *et seq.* Clarendon Press, Oxford (1954).

34. M. BORN, *Nachr. Akad. Wiss. Göttingen, II Math.–Physik. Kl.*, 1 (1951).

35. M. BORN AND K. HUANG, *Dynamical Theory of Crystal Lattices*, Appendix VIII, p. 406. Clarendon Press, Oxford (1954).

36. W. SHOCKLEY, *Electrons and Holes in Semiconductors*, p. 520 *et seq.* Van Nostrand, Princeton, N.J. (1950).

37. C. H. HENRY AND J. J. HOPFIELD, *Phys. Rev. Lett.* **15**, 964 (1965).

38. S. P. S. PORTO, B. TELL, AND T. C. DAMEN, *Phys. Rev. Lett.* **16**, 450 (1966).

39. M. BORN AND K. HUANG, *Dynamical Theory of Crystal Lattices*, p. 89 *et seq.* Clarendon Press, Oxford (1954).

40. L. MERTIN, *Z. Naturforsch.* **A15**, 47 (1960).

41. C. K. ASAWA, *Phys. Rev.* **B2**, 2068 (1970).

42. M. BORN AND K. HUANG, *Dynamical Theory of Crystal Lattices*, Appendix VI, p. 398. Clarendon Press, Oxford (1954).

43. H. A. LORENTZ, *Theory of Electrons*, pp. 137–150. Teubner, Leipzig (1909).

44. B. SZIGETI, *Trans. Faraday Soc.* **45**, 155 (1949).

45. R. J. HARDY AND A. M. KARO, *Phys. Rev.* **B 7**, 4696 (1973).

46. B. G. DICK AND A. W. OVERHAUSER, *Phys. Rev.* **112**, 90 (1958).

47. J. E. HANLON AND A. W. LAWSON, *Phys. Rev.* **113**, 472 (1959).

48. B. SZIGETI, *Proc. Roy. Soc. (London)* **A204**, 51 (1950).

49. R. A. COWLEY, *Adv. Phys.* **12**, 421 (1963).

50. H. BILZ, in *Phonons in Perfect Lattices and Lattices with Point Imperfections* (R. W. H. Stevenson, ed.), p. 208. Plenum, New York (1966).
51. D. J. MONTGOMERY AND R. J. MISHO, *Nature (London)* **183**, 103 (1959).
52. W. KOHN, *Phys. Rev. Lett.* **2**, 393 (1959).
53. K. B. TOLPYGO, *Zh. Eksp. Theor. Fiz.* **20**, 497 (1950).
54. K. B. TOLPYGO AND I. G. ZASLAVSKAYA, *Ukr. Fiz. Zh.* **1**, 226 (1956).
55. A. A. DEMIDENKO, Z. A. DEMIDENKO, AND K. B. TOLPYGO, *Ukr. Fiz. Zh.* **3**, 741 (1956).
56. Z. A. DEMIDENKO AND K. B. TOLPYGO, *Fiz. Tverd. Tela* **3**, 3435 (1961). Translation: *Sov. Phys.-Solid State* **3**, 2493 (1962).
57. V. S. MASHKEVITCH AND K. B. TOLPYGO, *Dokl. Akad. Nauk SSSR* **111**, 575 (1956). Translation: *Sov. Phys.-Dokl.* **1**, 690 (1957).
58. V. S. MASHKEVITCH AND K. B. TOLPYGO, *Zh. Eksp. Teor. Fiz.* **32**, 520 (1957). Translation: *Sov. Phys.-JETP* **5**, 435 (1957).
59. V. S. MASHKEVITCH, *Zh. Eksp. Theor. Fiz.* **36**, 1736 (1959). Translation: *Sov. Phys.-JETP* **9**, 1237 (1959).
60. K. B. TOLPYGO, *Fiz. Tverd. Tela* **3**, 943 (1961). Translation: *Sov. Phys.-Solid State* **3**, 185 (1961).
61. Z. A. DEMIDENKO, T. I. KUCHER, AND K. B. TOLPYGO, *Fiz. Tverd. Tela* **3**, 2482 (1961). Translation: *Sov. Phys.-Solid State* **3**, 1803 (1962).
62. Z. A. DEMIDENKO, T. I. KUCHER, AND K. B. TOLPYGO, *Fiz. Tverd. Tela* **4**, 104 (1962). Translation: *Sov. Phys.-Solid State* **4**, 73 (1962).
63. K. B. TOLPYGO, *Fiz. Tverd. Tela* **2**, 2655 (1960). Translation: *Sov. Phys.-Solid State* **2**, 2367 (1961).
64. V. S. MASHKEVITCH, *Fiz. Tverd. Tela* **2**, 2629 (1960). Translation: *Sov. Phys.-Solid State* **2**, 2345 (1961).
65. J. YAMASHITA AND T. KUROSAWA, *J. Phys. Soc. Japan* **10**, 610 (1955).
66. G. H. WANNIER, *Phys. Rev.* **52**, 191 (1937).
67. A. D. B. WOODS, W. COCHRAN, AND B. N. BROCKHOUSE, *Phys. Rev.* **119**, 980 (1960).
68. A. D. B. WOODS, B. N. BROCKHOUSE, R. A. COWLEY, AND W. COCHRAN, *Phys. Rev.* **131**, 1025 (1963).
69. R. A. COWLEY, W. COCHRAN, B. N. BROCKHOUSE, AND A. D. B. WOODS, *Phys. Rev.* **131**, 1030 (1963).
70. R. A. COWLEY, *Proc. Roy. Soc. (London)* **A268**, 109, 121 (1962).
71. S. K. SINHA, *CRC Crit. Rev. Solid State Sci.* **3**, 273 (1974).
72. J. R. HARDY, *Phil. Mag.* **4**, 1278 (1959).
73. J. R. HARDY, *Phil. Mag.* **7**, 315 (1962).
74. J. R. HARDY AND A. M. KARO, *Phil. Mag.* **5**, 859 (1960).
75. A. M. KARO AND J. R. HARDY, *Phys. Rev.* **129**, 2024 (1963).
76. W. COCHRAN, *Proc. Roy. Soc. (London)* **A253**, 260 (1959).
77. U. SCHRÖDER, *Solid State Commun.* **4**, 347 (1966).
78. V. NÜSSLEIN AND U. SCHRÖDER, *Phys. Status Solidi* **21**, 309 (1967).

79. M. J. L. SANGSTER, G. PECKHAM, AND D. H. SAUNDERSON, *J. Phys. C.* **3**, 1026 (1970).

80. W. COCHRAN, *CRC Crit. Rev. Solid State Sci.* **2**, 1 (1971).

81. J. R. TESSMAN, A. H. KAHN, AND W. SHOCKLEY, *Phys. Rev.* **92**, 890 (1953).

82. J. PIRENNE AND E. KARTHEUSER, *Physica* **30**, 2005 (1964).

83. S. S. JASWAL AND T. P. SHARMA, *J. Phys. Chem. Solids* **34**, 509 (1973).

84. S. L. ADLER, *Phys. Rev.* **126**, 413 (1961).

85. N. WISER, *Phys. Rev.* **129**, 62 (1963).

86. S. K. SINHA, *Phys. Rev.* **169**, 477 (1968); **177**, 1256 (1969).

87. S. K. SINHA, R. P. GUPTA, AND D. L. PRICE, *Phys. Rev. Lett.* **26**, 1324 (1971); *Phys. Rev.* **B 9**, 2564 (1974).

88. D. L. PRICE, S. K. SINHA, AND R. P. GUPTA, *Phys. Rev.* **B 9**, 2573 (1974).

89. M. P. VERMA AND R. K. SINGH, *Phys. Status Solidi* **33**, 769 (1969).

90. R. K. SINGH AND M. P. VERMA, *Phys. Status Solidi* **36**, 335 (1969); **38**, 851 (1970).

91. S. O. LUNDQVIST, *Ark. Fysik* **6**, 25 (1952); **9**, 435 (1955); **12**, 263 (1957).

92. A. A. MARADUDIN, private communication.

93. S. L. CUNNINGHAM, Ph.D. thesis, University of Nebraska (1971).

94. E. R. LEVIN AND E. L. OFFENBACHER, *Phys. Rev.* **118**, 1142 (1960).

95. J. YAMASHITA, *Prog. Theor. Phys.* (*Kyoto*) **8**, 280 (1952); **12**, 454 (1954).

96. J. R. HARDY AND A. B. LIDIARD, *Phil. Mag.* **15**, 825 (1967).

97. J. R. HARDY, *Phil. Mag.* **6**, 27 (1961).

98. A. HERPIN, *J. Phys. Radium* **14**, 611 (1953).

99. J. LOTHE, *Arch. Math. Naturvid.* **55**, 20 (1959).

100. B. G. DICK, *Phys. Rev.* **129**, 1583 (1963).

101. G. BENEDEK AND A. A. MARADUDIN, *J. Phys. Chem. Solids* **29**, 423 (1968).

102. P. P. EWALD, *Ann. Physik* (*Leipzig*) **54**, 519, 557 (1917).

103. P. P. EWALD, *Ann. Physik* (*Leipzig*) **64**, 253 (1921).

104. A. A. MARADUDIN, E. W. MONTROLL, G. H. WEISS, AND I. P. IPATOVA, *Theory of Lattice Dynamics in the Harmonic Approximation*, 2nd ed., p. 520 *et seq.* Academic Press, New York (1971).

105. S. Y. TONG AND A. A. MARADUDIN, *Phys. Rev.* **181**, 1318 (1969).

106. G. GILAT AND G. DOLLING, *Phys. Lett.* **8**, 304 (1964).

107. G. GILAT AND L. J. RAUBENHEIMER, *Phys. Rev.* **144**, 390 (1966).

108. L. J. RAUBENHEIMER AND G. GILAT, *Phys. Rev.* **157**, 586 (1967).

109. A. M. KARO AND J. R. HARDY, *Phys. Rev.* **141**, 696 (1966).

110. L. VAN HOVE, *Phys. Rev.* **89**, 1189 (1953).

111. J. C. PHILLIPS, *Phys. Rev.* **104**, 1263 (1956).

112. A. A. MARADUDIN, E. W. MONTROLL, G. H. WEISS, AND I. P. IPATOVA, *Theory of Lattice Dynamics in the Harmonic Approximation*, 2nd ed., p. 429 *et seq.* Academic Press, New York (1971).

113. C. M. EISENHAUER, I. PELAH, D. J. HUGHES, AND H. PALEVSKY, *Phys. Rev.* **109**, 1046 (1958).
114. S. S. JASWAL, *Phys. Rev.* **144**, 428 (1966).
115. A. M. KARO AND J. R. HARDY, *Phys. Rev.* **181**, 1272 (1969).
116. M. MERISALO, *Ann. Akad. Sci. Fennicae* **245**, 1 (1967).
117. P. G. DAWBER AND R. J. ELLIOTT, *Proc. Roy. Soc. (London)* **A273**, 222 (1963).
118. T. H. K. BARRON, A. J. LEADBETTER, J. A. MORRISON, AND L. S. SALTER, in *Inelastic Scattering of Neutrons by Solids and Liquids*, p. 49. International Atomic Energy Agency, Vienna (1963).
119. A. EINSTEIN, *Ann. Physik* **22**, 180 (1907).
120. M. PLANCK, *Ann. Physik* **4**, 553 (1901).
121. P. DEBYE, *Ann. Physik* **39**, 789 (1912).
122. P. DEBYE, in *Lattice Dynamics* (R. D. Wallis, ed.), p. 9. Pergamon, Oxford (1965).
123. M. BLACKMAN, *Proc. Roy. Soc. (London)* **A148**, 365, 385 (1935).
124. M. BLACKMAN, *Proc. Roy. Soc. (London)* **A149**, 117 (1935).
125. M. BLACKMAN, *Phil. Trans. Roy. Soc. (London)* **A236**, 103 (1936).
126. M. BLACKMAN, *Proc. Roy. Soc. (London)* **A159**, 416 (1937).
127. M. BLACKMAN, *Proc. Camb. Phil. Soc.* **33**, 94 (1937).
128. T. H. K. BARRON AND M. L. KLEIN, *Phys. Rev.* **127**, 1997 (1962).
129. E. W. KELLERMANN, *Proc. Roy. Soc. (London)* **A178**, 17 (1941).
130. W. T. BERG AND J. A. MORRISON, *Proc. Roy. Soc. (London)* **A242**, 467 (1957). See also J. A. MORRISON, D. PATTERSON, AND J. S. DUGDALE, *Can. J. Chem.* **33**, 375 (1955); J. A. MORRISON AND D. PATTERSON, *Trans. Faraday Soc.* **55**, 764 (1956).
131. T. H. K. BARRON, W. T. BERG, AND J. A. MORRISON, *Proc. Roy. Soc. (London)* **A242**, 478 (1957).
132. M. SORAI, Ph.D. thesis, University of Osaka (1967).
133. G. LIEBFRIED AND W. LUDWIG, in *Solid State Physics* (F. Seitz and D. Turnbull, eds.), Vol. 12, p. 275. Academic Press, New York (1961).
134. R. P. LOWNDES AND D. H. MARTIN, *Proc. Roy. Soc. (London)* **A308**, 473 (1969).
135. S. S. JASWAL AND J. R. HARDY, *Phys. Rev.* **171**, 1090 (1968).
136. J. L. VERBLE, J. L. WARREN, AND J. L. YARNELL, *Phys. Rev.* **168**, 980 (1968).
137. W. M. LOMER AND G. G. LOW, in *Thermal Neutron Scattering* (P. A. Egelstaff, ed.), p. 1. Academic Press, New York (1965).
138. T. SMITH, in *Phonons in Perfect Lattices and Lattices with Point Imperfections* (R. W. H. Stevenson, ed.), p. 161. Plenum, New York (1966).
139. W. J. L. BUYERS AND T. SMITH, *Phys. Rev.* **150**, 758 (1966).
140. J. S. MELVIN, J. D. PIRIE, AND T. SMITH, *Phys. Rev.* **175**, 1082 (1968).
141. W. J. L. BUYERS, J. D. PIRIE, AND T. SMITH, *Phys. Rev.* **165**, 999 (1968).

142. J. D. PIRIE AND T. SMITH, *J. Phys. C* **1**, 648 (1968).
143. J. E. ELDRIDGE AND T. R. LOMER, *Proc. Phys. Soc.* **91**, 459 (1967).
144. J. F. VETELINO, S. S. MITRA, AND V. K. NAMJOSHI, *Phys. Rev.* **B 2**, 2167 (1970).
145. G. DOLLING, H. G. SMITH, R. M. NICKLOW, P. R. VIJAYARAGHAVAN, AND M. K. WILKINSON, *Phys. Rev.* **168**, 970 (1968).
146. B. N. BROCKHOUSE AND P. K. IYENGAR, *Phys. Rev.* **111**, 747 (1958).
147. W. COCHRAN, *CRC Crit. Rev. Solid State Sci.* **2**, 40 (1971).
148. R. A. COWLEY, in *Lattice Dynamics* (R. F. Wallis, ed.), p. 295. Pergamon, Oxford (1965).
149. R. F. WALLIS AND A. A. MARADUDIN, *Phys. Rev.* **125**, 1277 (1962).
150. A. A. MARADUDIN AND R. F. WALLIS, *Phys. Rev.* **123**, 777 (1961).
151. J. E. ELDRIDGE, *Phys. Rev.* **B 6**, 1510 (1972).
152. J. E. ELDRIDGE AND R. HOWARD, *Phys. Rev.* **B 7**, 4652 (1973).
153. M. LAX AND E. BURSTEIN, *Phys. Rev.* **97**, 39 (1955).
154. R. A. COWLEY, in *Phonons in Perfect Lattices and Lattices with Point Imperfections* (R. W. H. Stevenson, ed.), p. 170. Plenum, New York (1966).
155. B. SZIGETI, *Proc. Roy. Soc. (London)* **A252**, 217 (1959).
156. B. SZIGETI, *Proc. Roy. Soc. (London)* **A261**, 274 (1961).
157. R. LOUDON, *Phys. Rev.* **137**, A1784 (1965).
158. C. SMART, G. R. WILKINSON, A. M. KARO, AND J. R. HARDY, in *Lattice Dynamics* (R. F. Wallis, ed.), p. 387. Pergamon, Oxford (1965).
159. F. A. JOHNSON, *Proc. Phys. Soc.* **73**, 265 (1959).
160. B. FRITZ, U. GRÖSS, AND D. BÄUERLE, *Phys. Status Solidi* **11**, 231 (1965).
161. H. BILZ, D. STRAUCH, AND B. FRITZ, *J. de Phys. Colloque* **C2-3** (1968).
162. M. V. KLEIN, in *The Physics of Color Centers* (W. Beall Fowler, ed.), p. 429. Academic Press, New York (1968).
163. R. S. KRISHNAN, in *Lattice Dynamics* (R. F. Wallis, ed.), p. 429. Pergamon, Oxford (1965).
164. R. LOUDON, *Proc. Roy Soc. (London)* **A275**, 218 (1963).
165. M. BORN AND M. BRADBURN, *Proc. Roy. Soc. (London)* **A241**, 105 (1948).
166. M. BORN AND K. HUANG, *Dynamical Theory of Crystal Lattices*, p. 367 *et seq.* Clarendon Press, Oxford (1954).
167. M. KRAUZMAN, *Compt. Rend. Acad. Sci.* **B265**, 1029 (1967).
168. M. KRAUZMAN, *Compt. Rend. Acad. Sci.* **B266**, 186 (1968).
169. S. L. CUNNINGHAM, J. R. HARDY, AND M. HASS, in *Proceedings of the Second International Conference on Light Scattering in Solids* (M. Balkanski, ed.), p. 257. Flammarion, Paris (1971).
170. J. R. HARDY, A. M. KARO, I. W. MORRISON, C. T. SENNETT, AND J. P. RUSSEL, *Phys. Rev.* **179**, 837 (1969).
171. R. LOUDON, *Adv. Phys.* **13**, 423 (1964).
172. R. A. COWLEY, *Proc. Phys. Soc.* **84**, 281 (1964).
173. A. D. BRUCE AND R. A. COWLEY, *J. Phys. C* **5**, 595 (1972).

174. A. D. BRUCE, *J. Phys. C* **5**, 2909 (1972).
175. G. SCHÄFER, *J. Phys. Chem. Solids* **12**, 233 (1960).
176. B. FRITZ, *J. Phys. Chem. Solids* **23**, 375 (1962).
177. T. TIMUSK AND M. V. KLEIN, *Phys. Rev.* **141**, 664 (1966).
178. I. M. LIFSHITZ, *J. Phys. USSR* **7**, 215 (1943).
179. I. M. LIFSHITZ, *J. Phys. USSR* **7**, 249 (1943).
180. I. M. LIFSHITZ, *J. Phys. USSR* **8**, 89 (1944).
181. P. G. DAWBER AND R. J. ELLIOTT, *Proc. Phys. Soc.* **81**, 453 (1963).
182. R. S. LEIGH AND B. SZIGETI, *Proc. Roy. Soc. (London)* **A301**, 211 (1967).
183. R. S. LEIGH AND B. SZIGETI, *Phys. Rev. Lett.* **19**, 566 (1967).
184. R. C. NEWMAN AND J. B. WILLIS, *J. Phys. Chem. Solids* **26**, 373 (1965).
185. R. J. ELLIOTT AND D. W. TAYLOR, *Proc. Roy. Soc. (London)* **A296**, 161 (1967).
186. C. W. McCOMBIE, in *Optical Properties of Solids* (S. S. Mitra and S. Nudelman, eds.), p. 595. Plenum, New York (1969).
187. C. W. McCOMBIE, in *Far Infra-Red Properties of Solids* (S. S. Mitra and S. Nudelman, eds.), p. 297. Plenum, New York (1970).
188. G. B. WRIGHT, ed., *Light Scattering Spectra of Solids*, p. 439 *et seq.* Springer-Verlag, New York (1969).
189. R. T. HARLEY, J. B. PAGE, AND C. T. WALKER, *Phys. Rev.* **133**, 1365 (1971), and references cited therein.
190. C. W. McCOMBIE AND J. SLATER, *Proc. Phys. Soc.* **84**, 499 (1964).
191. R. J. ELLIOTT AND D. W. TAYLOR, *Proc. Phys. Soc.* **83**, 189 (1964).
192. R. F. CALDWELL AND M. V. KLEIN, *Phys. Rev.* **158**, 851 (1967), and references cited therein.
193. P. G. KLEMENS, in *Solid State Physics* (F. Seitz and D. Turnbull, eds.), Vol. 7, p. 1. Academic Press, New York (1958).
194. A. A. MARADUDIN, in *Solid State Physics* (F. Seitz and D. Turnbull, eds.), Vol. 18, p. 273, and Vol. 19, p. 1. Academic Press, New York (1966 and 1967).
195. A. M. KARO AND J. R. HARDY, *Phys. Rev.* **B 12**, 690 (1975).
196. C. W. McCOMBIE, J. A. D. MATTHEW, AND A. M. MURRAY, *J. Appl. Phys. (Suppl.)* **33**, 359 (1962).
197. T. P. DAS AND E. L. HAHN, *Solid State Physics*, Suppl. 1. Academic Press, New York (1958).
198. N. F. MOTT AND M. J. LITTLETON, *Trans. Faraday Soc.* **34**, 485 (1938).
199. R. A. JOHNSON, *J. Phys. F* **3**, 295 (1973).
200. E. S. RITTNER, R. A. HUTNER, AND F. K. DU PRÉ, *J. Chem. Phys.* **17**, 198 (1949).
201. E. S. RITTNER, R. A. HUTNER, AND F. K. DU PRÉ, *J. Chem. Phys.* **17**, 204 (1949).
202. I. D. FAUX AND A. B. LIDIARD, *Z. Naturforsch.* **26**, 62 (1971).
203. I. M. BOSWARVA AND A. B. LIDIARD, *Phil. Mag.* **16**, 805 (1967).
204. P. BRAUER, *Z. Naturforsch.* **6a**, 561, 562 (1959).

205. A. SCHOLZ, *Phys. Status Solidi* **25**, 285 (1968).

206. H. J. KANZAKI, *J. Phys. Chem. Solids* **2**, 24 (1957).

207. H. J. KANZAKI, *J. Phys. Chem. Solids* **2**, 37 (1957).

208. J. R. HARDY, *J. Phys. Chem. Solids* **15**, 39 (1960).

209. J. R. HARDY, *J. Phys. Chem. Solids*, **23**, 113 (1962).

210. P. D. SCHULZE AND J. R. HARDY, *Phys. Rev.* **B 5**, 3270 (1972).

211. P. D. SCHULZE AND J. R. HARDY, *Phys. Rev.* **B 6**, 1580 (1972).

212. P. D. SCHULZE, Ph.D. thesis, University of Nebraska (1971).

213. W. BIERMANN, *Z. Phys. Chem.* (*Frankfurt*) **25**, 90 (1960).

214. W. BIERMANN, *Z. Phys. Chem.* (*Frankfurt*) **25**, 253 (1960).

215. M. BEYELER AND D. LAZARUS, *Solid State Commun.* **7**, 1487 (1969).

216. D. LAZARUS, D. N. YOON, AND R. N. JEFFERY, *Z. Naturforsch.* **26a**, 56 (1971).

217. H. L. HEINISCH AND S. S. JASWAL, *Phys. Rev.* **B 9**, 2754 (1974).

218. O. K. ONG AND J. M. VAIL, *Phys. Rev.* **B 8**, 1636 (1973).

219. J. D. ESHELBY, *J. Appl. Phys.* **25**, 255 (1954).

220. J. R. HARDY, *J. Phys. Chem. Solids* **29**, 2009 (1968).

221. A. A. MARADUDIN, *J. Phys. Chem. Solids* **9**, 1 (1958).

222. L. L. BOYER AND J. R. HARDY, *Phil. Mag.* **24**, 647 (1971).

223. L. L. BOYER AND J. R. HARDY, *Phil. Mag.* **26**, 225 (1972).

224. R. S. LEIGH, B. SZIGETI, AND V. K. TEWARY, *Proc. Roy. Soc.* (*London*) **A320**, 505 (1970).

225. W. COCHRAN, *Acta Cryst.* **A27**, 556 (1971).

Index